Studies in Geometry Series

Circles
Workbook

Relationships with Segments and Angles

Tammy Pelli

A Breath of Fresh Air

GarlicPress

Published by:
Garlic Press
605 Powers St.
Eugene, OR 97402

ISBN 978-1-9308-2045-6
Order Number GP-145
Printed in China

www.garlicpress.com

Table of Contents

Introduction

- The purpose of this workbook is to explore the relationships that exist between lines, angles, arcs, and circles. The fundamentals of geometry are required to develop the specific applications to circles – reminders of the basics are given, as needed, in the text, but the student does need to be familiar with the basics of lines, segments, angles, triangles, etc.

- Because every textbook is different, reference to specific theorems and postulates is not made overtly so that this workbook can complement a student's study of geometry without conflicting with his/her school work. Many theorems, postulates, and definitions are presented informally and their applications are developed in the explanation and the practice exercises.

- This book requires basic algebra skills. Algebra is used as a method of explanation for some of the ideas presented. Additionally, algebraic examples are integrated into the practice exercises.

- Towards the end of the workbook, you will find an Exam so that you can check to find if you have mastered the concepts presented here.

- A brief Glossary of vocabulary and basic concepts is found at the end of the workbook, preceding the Answer Key.

- The Answer Key provides the answers to all practice exercises. In many cases, it also provides a thorough explanation of the thinking involved in specific problems.

Chapter 1

Circles and Segments

Circle Parts Defined

○ Informally, we know what a circle is – it's round, like the sun or a tire. Many objects from everyday life are in the shape of a circle.

○ **In geometry, a circle has a precise definition: A <u>circle</u> is a collection of points equidistant from a single point. That single point is called the <u>center</u> of the circle.**

○ **Note: The circle is only the points (the line you draw to make the circle).** The interior of the circle is not part of the circle – see Circle *I*: point *Z* is not <u>on the circle</u>, it is <u>in the interior of the circle</u>. Most of the everyday shapes we think of as being circular are really discs, not circles. A circle is the edge, like the rim on the top of a soda can, and doesn't include the interior (like where you put your food on the plate).

○ **Naming Circles:** Circles are named for their centers. Circle *I* is noted as ⊙*I*.

○ The **diameter** of ⊙*I* is $\overline{EG}$. It is a chord which passes through the center of the circle.

○ A **chord** is any segment in a circle with endpoints on the circle. A diameter is the longest possible chord in a circle.

○ A **radius** is a segment that goes from the center of a circle to a point on the circle. There are three radii drawn in ⊙*I*: $\overline{EI}$, $\overline{IG}$, and $\overline{JI}$. A diameter contains two radii.

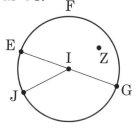

Circle *I*

○ A **secant** is a line segment or ray which intersects a circle at two points. $\overleftrightarrow{ZW}$ of ⊙*X* is a secant of ⊙*X*.

○ In ⊙X, $\overline{ZW}$ and $\overline{VU}$ are **chords** because their endpoints are on the circle. All chords are pieces of secants, but they do not continue outside the circle as secants can.

○ A **tangent** intersects a circle at only one point. $\overleftrightarrow{YT}$ is a tangent line. *Y* is the **point of tangency**.

○ Radius $\overline{XY}$ is perpendicular (⊥) to tangent $\overleftrightarrow{YT}$. A radius is always perpendicular to a tangent at the point of tangency.

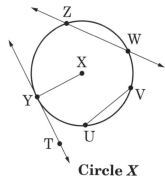

Circle X

Example:
Identifying parts of circles
Name: ⊙*E*
Chords: $\overline{DS}$, $\overline{JB}$
Tangent: $\overleftrightarrow{GT}$
Right angles: ∠*GJE* and ∠*TJE*
Point of tangency: *J*

Diameter: $\overline{JB}$ (longest chord)
Radii: $\overline{JE}$, $\overline{EB}$, $\overline{EL}$
Secant: $\overleftrightarrow{DS}$, $\overline{JB}$

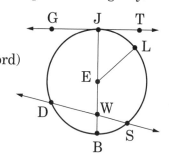

Practice

Identify any circles, chords, tangents, right angles, tangencies, diameters, radii or secants.

1. Identify the parts of $\odot C$.

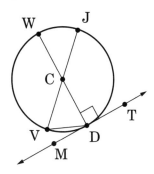

2. Identify the parts of $\odot S$.

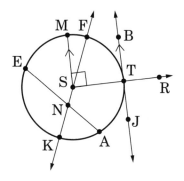

3. Identify the parts of $\odot B$ and $\odot D$.

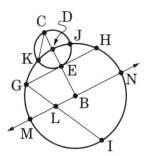

4. Identify the parts of $\odot L$.

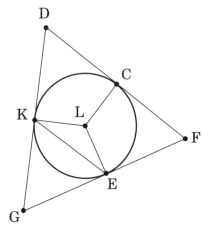

5. $\odot W$ and $\odot A$ are internally tangent. Internally tangent circles intersect at exactly one point and one circle is in the interior of the other circle. Identify the parts of each.

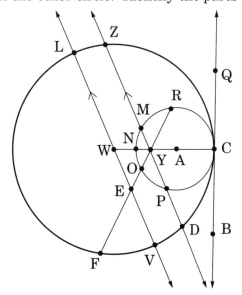

6. These are concentric circles. They both have M as their center. The smaller circle is $\odot M_1$ and the larger circle is $\odot M_2$. Identify the parts of each.

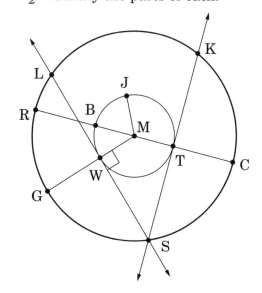

Diameter and Radius

○ ***The length of a radius is equal to one half the length of a diameter*** – *or a diameter is twice as long as a radius of the same circle.*

Since $AB + BD = AD$ (segment addition)
and $AB = BD$, by substitution,
$AB + AB = AD$, and $2AB = AD$.

○ **Congruent Circles**: Two circles are congruent if their radii are congruent.

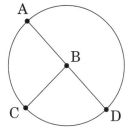

$\overline{AD}$ is a diameter
$AD = 4cm$
$\overline{BC}$, $\overline{BD}$ and $\overline{BA}$ are radii
$BC = BD = BA = 2cm$

Practice

Based on the information in each figure, calculate the measurements requested.
Refer to page 3 for definitions.

1. $\odot N$.

    $ON =$

    $NM =$

    $OM =$

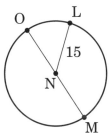

2. $\odot B$ is internally tangent to $\odot A$.

    $DC = 6$

    $AD =$

    $AC =$

    $AB =$

    $BC =$

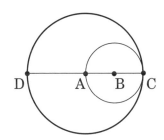

3. $\overleftrightarrow{ZX}$ is tangent to $\odot V$. $VW = 8$, $ZX = 3$.

    $VX =$

    $YX =$

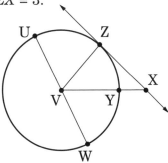

4. $\odot E$ and $\odot F$ are externally tangent because they intersect only in G and they do not share any interior space. G and H are points of tangency. $HI = 16$, $GF = 9$.

    $EH =$

    $HK =$

    $HF =$

    $EF =$

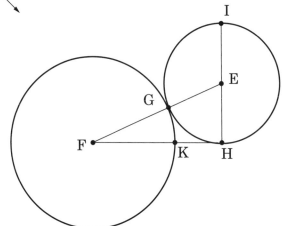

Chapter 2

Arcs and Angles

Arcs and Central Angles

○ ∠*ABC* has the center of the circle as the vertex of its angle. That makes ∠*ABC* a **central angle**.

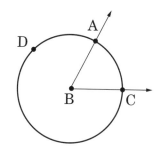

○ Central angles intersect the circle, forming **arcs**. An **arc** is a part of the circle. Arcs on ⊙*B* include $\overarc{AC}$, $\overarc{ADC}$, $\overarc{AD}$, and $\overarc{DC}$.

There are 3 classifications of arcs

○ A **semicircle** is half of a circle. The diameter cuts a circle into two equal parts – each is a semicircle. $\overarc{EFG}$ and $\overarc{EHG}$ are semicircles. Semicircles are named for their endpoints and one other point on the arc.

○ A **minor arc** is any arc which is smaller than a semicircle. $\overarc{EF}$, $\overarc{FG}$, $\overarc{EH}$, and $\overarc{HG}$ are minor arcs in ⊙*I*. Minor arcs are named for their endpoints.

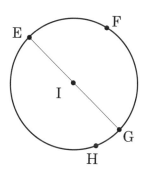

○ A **major arc** is any arc which is larger than a semicircle. Like semicircles, major arcs are named for their endpoints and one other point on the arc. $\overarc{EFH}$ is the same arc as $\overarc{EGH}$, because they have the same endpoints and both *F* and *G* are on the same arc. $\overarc{EH}$ is the minor arc that completes the circle. $\overarc{EHF}$ is the same arc as $\overarc{EGF}$. $\overarc{FEH}$ is also a major arc.

Minor arc $\overarc{MN}$ has been darkened on ⊙*L*.
Major arc $\overarc{MON}$ completes the circle.

Practice

For each circle list all minor arcs, major arcs, semicircles and central angles.

1.

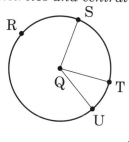

2.

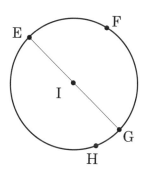

3.

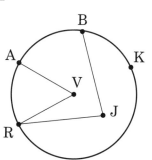

4.

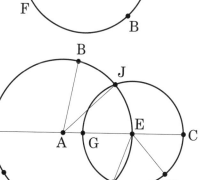

M e a s u r e s   o f   A r c s   a n d   C e n t r a l   A n g l e s

○ An **intercepted arc** is formed when a central angle intersects a circle. *The measurement of the intercepted arc is the same as the measurement of the central angle which defines it.*

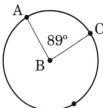

$m\angle ABC = 89°$

so, $m\overarc{AC} = 89°$

Examples:

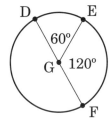

1. $m\angle DGE = 60°$, so $m\overarc{DE} = 60°$
   $m\angle EGF = 120°$, so $m\overarc{EF} = 120°$

2. The measure of any semicircle is 180.°
   $m\angle DGF = 180°$, so $m\overarc{DEF} = 180°$

○ **Adjacent arcs** share an endpoint but do not overlap. For example: $\overarc{YU}$ is adjacent to $\overarc{UV}$, $\overarc{UW}$ is adjacent to $\overarc{WY}$, $\overarc{XY}$ is adjacent to $\overarc{VX}$.

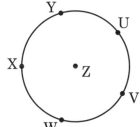

○ **Nonadjacent arcs** do not share an endpoint. For example: $\overarc{YX}$ and $\overarc{WU}$, $\overarc{YU}$ and $\overarc{WV}$ are non-adjacent arcs.

○ **Arc Addition**: Adjacent arcs can be added together the same way segments can be added for segment addition.

Examples:

1. Find the measure of $\overarc{ABC}$.
   $m\overarc{ABC} = m\overarc{BA} + m\overarc{BC} = m\angle AEB + m\angle BEC$
   $= 70° + 146° = 216°$

2. Find the measure of $\overarc{BDC}$.
   $m\overarc{BDC} = m\overarc{BA} + m\overarc{AD} + m\overarc{DC}$
   $= m\angle BEA + m\angle AED + m\angle DEC$
   $= 70° + 48° + 96° = 214°$

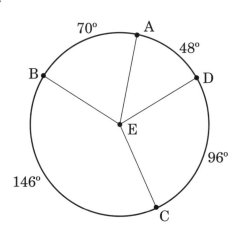

○ **Congruent Arcs**: Two arcs are congruent if . . .
1.   they have the same measure,
2.   they are in the same circle or in congruent circles.

Examples:

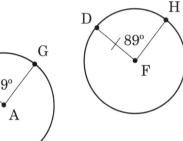

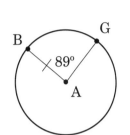

1. $m \angle BAG = m \angle DFH = 89°$ so m $\widehat{BG}$ = m $\widehat{DH}$
$\overline{BA} \cong \overline{AG} \cong \overline{DF} \cong \overline{FH}$ so the circles are
congruent. Therefore, $\widehat{BG} \cong \widehat{DH}$.

2. $m \angle CAB = m \angle BAD$, so m $\widehat{BC}$ = m $\widehat{BD}$
The arcs are in the same circle.
Therefore $\widehat{BC} \cong \widehat{BD}$.

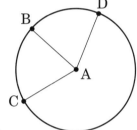

$m \angle CAB = 71°$
$m \angle BAD = 71°$

3. $m \angle BDC = m \angle FEG$
However, $\widehat{DB} \neq \widehat{EF}$, so $\widehat{BC} \neq \widehat{FG}$.

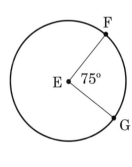

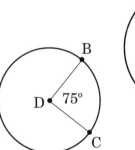

Practice:

1. $m\overset{\frown}{AC} =$

   $m\overset{\frown}{ADC} =$

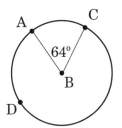

2. $m\overset{\frown}{YL} =$

   $m\overset{\frown}{YLM} =$

   $m \angle LXM =$

   $m \angle YXM =$

   $m \angle ZXY =$

   $m \angle ZXL =$

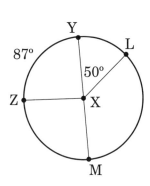

3. $m \angle ANV =$

   $m \angle AND =$

   $m\overset{\frown}{AD} =$

   $m\overset{\frown}{DQV} =$

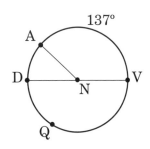

4. $m\overset{\frown}{EG} = 75°$ $m\overset{\frown}{FH} = 65°$ $m\overset{\frown}{EH} = 245°$

   $m\overset{\frown}{FG} =$

   $m \angle EJF =$

   $m \angle GJH =$

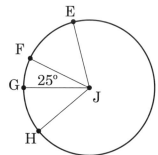

5. Name a pair of adjacent arcs.

   Name a pair of nonadjacent arcs.

   Name a major arc.

   Name a minor arc.

   Name an acute angle.

   Name an obtuse angle.

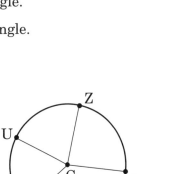

6. $m\overset{\frown}{CT} = 35°$ $m \angle CRH = 80º$

   $m \angle TRL =$

   $m\overset{\frown}{CL} =$

   $m\overset{\frown}{LB} =$

   $m \angle CRB =$

   $m\overset{\frown}{CHB} =$

   $m\overset{\frown}{HB} =$

   $m \angle HRB =$

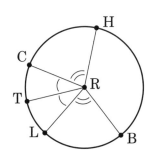

Chapter 3

Chords

Chords Defined

○ A **chord** is a segment which has its endpoints on a circle.

○ A **secant** is a line, ray or segment that intercepts a circle at two points.

○ The relationship between arcs and chords is similar to the relationship of arcs and central angles.

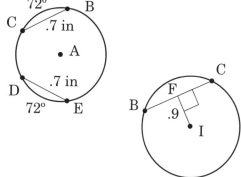

• If two chords in a circle are congruent, then the arcs bounded by their endpoints are congruent.

Since $CB = DE = .7$ inches, $\overset{\frown}{CB} \cong \overset{\frown}{DE}$.

• If two arcs in a circle are congruent, then the chords connecting their endpoints are congruent.

Since $m\overset{\frown}{CB} = 72° = m\overset{\frown}{DE}$, $\overline{CB} \cong \overline{DE}$.

○ To measure the distance from a chord to the center of a circle, measure along a line through the center which is perpendicular to the chord. For example, the chord in $\odot I$ is .9cm from the center.

○ There are also special relationships between radii, diameters, and chords in circles.

Diameter – Chord Relationship

$m\overset{\frown}{GB} = 56°$

$m\overset{\frown}{CG} = 56°$

$CI = 1.4$cm

$IB = 1.4$cm

If a diameter is perpendicular to a chord, then it bisects the chord and the arc defined by the endpoints of the chords.

The converse of this statement is also true.

Radius – Chord Relationship

$\overline{DK} \cong \overline{KW}$

$\overline{FE} \cong \overline{XJ}$

$\overline{FE} \perp \overline{LK}$

$\overline{XJ} \perp \overline{KG}$

Since the chords $\overline{FE}$ and $\overline{XJ}$ are equidistant from K, the center of the circle, they are congruent to each other.

The converse of this statement is also true.

Example:

Given information: $DA = 9$ and $AG = 9$, $\overline{AD} \perp \overline{BC}$, and $DC = 8$.

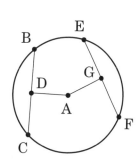

Conclusions: $DC = BD$, so $BD = 8$, and $BC = 16$. Since $\overline{BC}$ and $\overline{EF}$ are equidistant from the center of the circle, they are congruent. Therefore, $EF = 16$ and $EG = GF = 8$. Since $\overline{BC} \cong \overline{EF}$, their intercepted arcs are also congruent and $m\overset{\frown}{BC} = m\overset{\frown}{EF}$.

Practice

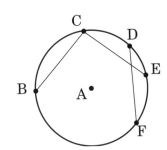

1. $BC = 22$ $FD =$ _____

    $CE = 22$ $\overline{BC} \cong$ _____ $\cong$ _____

    $\widehat{DF} \cong \widehat{CE}$ $m\widehat{CE} =$ _____ $=$ _____

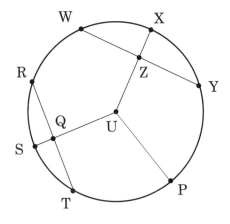

2. $SQ = 6$, $UP = 25$, $ZU = 19$.
    $\angle UZY$ and $\angle UQT$ are right angles.
    Name all congruent chords and arcs.

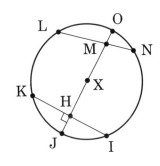

3. $m\widehat{JI} = 34°$, $m\angle KHJ = 90°$.

    $m\widehat{KI} =$

    $m\widehat{KJ} =$

    $m\widehat{KLI} =$

    Name any congruent chords.

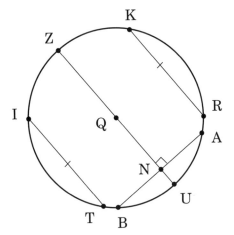

4. Name all congruent chords and arcs.
    How far is $\overline{AB}$ from Q if $ZU = 6$cm and $NU = 1$cm?

Chapter 4

Inscribed Figures

Inscribed Angles Defined

○ An **inscribed angle** is an angle in a circle with its vertex on the circle.
- ∠*RAQ* is inscribed on ⊙*B*.
- ∠*RBQ* is a central angle in ⊙*B*.

○ The measure of an inscribed angle is half the measure of the arc it intercepts.

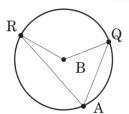

Examples:

1. $m\overarc{KY} = 62°$ $m\overarc{VD} = 74°$
   $m\angle KLY = 31°$ $m\angle VID = 37°$

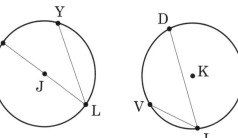

2. ∠*EFG* and ∠*EHG* are inscribed angles. Both of these angles intercept the same arc. Since the measure of an inscribed angle is half the measure of the intercepted arc, these 2 angles have the same measure.

   ∠*FEH* and ∠*HGF* are inscribed angles which intercept the same arc, $m\overarc{FH} = 56°$.
   $m\angle FEH = m\angle HGF = \frac{1}{2}m\overarc{FH} = 28°$

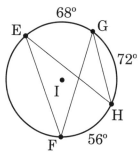

Practice

1. $m\overarc{BDC} = 206°$
   What is the measure of ∠*BDC*?

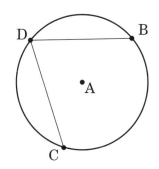

2. $m\overarc{NT} = 42°$
   What is the measure of ∠*NWT*?

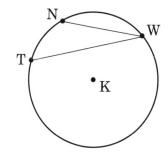

3. $m\overarc{PX} = 91°$ *and* $m\overarc{XE} = 47°$.
What is the measure of $\angle PBE$?

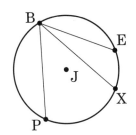

4. $m\overarc{GC} = 58°$ *and* $m\overarc{RG} = 81°$.
What is the measure of $\angle RGC$?

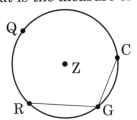

5. $m\angle BED =$

$m\angle CED =$

$m\overarc{CD} =$

$m\overarc{BD} =$

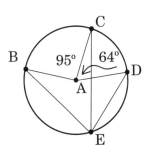

6. $m\overarc{NI} = m\overarc{JK} =$

$m\angle NMI =$

$m\angle NIM =$

$m\overarc{JK} =$

$m\angle JLK =$

$m\angle LJK =$

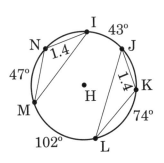

7. $m\angle YZW =$

$m\angle ZWY =$

$m\overarc{ZW} =$

$m\angle ZYW =$

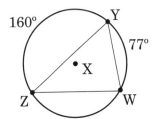

8. $m\angle ZIK =$

$m\angle KZI =$

$m\angle ZIJ =$

$m\angle KZJ =$

$m\angle ZJI =$

$m\angle ZKI =$

$m\angle ZKI + m\angle ZJI =$

$m\angle KZJ + m\angle KIJ =$

$m\angle IZJ =$

$m\angle KIJ =$

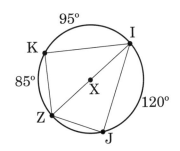

9. $m\overarc{CL} =$

$m\overarc{LV} =$

$m\angle LUV =$

$m\overarc{VU} =$

$m\overarc{UG} =$

$m\angle UVG =$

$m\angle KVG =$

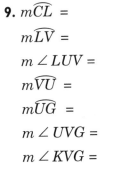

Inscribed Polygons

○ $\triangle ABC$ is **inscribed** in $\odot D$.

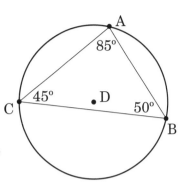

$\odot D$ is **circumscribed about** $\triangle ABC$.

Each vertex of $\triangle ABC$ is a point on $\odot D$.

$m \angle ACB = 45°$, so $m\widehat{AB} = 90°$.
$m \angle ABC = 50°$, so $m\widehat{AC} = 100°$.
$m \angle CAB = 85°$, so $m\widehat{CB} = 170°$.
$m \angle ABC + m \angle ACB + m \angle BAC = 180°$,
because it is a triangle.
$m\widehat{AC} + m\widehat{CB} + m\widehat{AB} = 360°$, because
there are 360° in a circle.

○ The irregular pentagon $VWXYZ$ is inscribed in $\odot Q$.

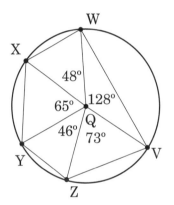

$m\widehat{WX} = m \angle WQX = 48°$ $m\widehat{XY} = m \angle XQY = 65°$

$m\widehat{YZ} = m \angle YQZ = 46°$ $m\widehat{ZV} = m \angle ZQV = 73°$

$m\widehat{VW} = m \angle VQW = 128°$

$m \angle WXY = \frac{1}{2}\left(m\widehat{WV} + m\widehat{VZ} + m\widehat{YZ}\right)$

$\qquad = \frac{1}{2}(128 + 73 + 46) = \frac{1}{2}(247) = 123.5°$

$m \angle YZV = \frac{1}{2}\left(m\widehat{XY} + m\widehat{XW} + m\widehat{WV}\right)$

$\qquad \frac{1}{2}(65 + 48 + 128) = \frac{1}{2}(241) = 120.5°$

○ A diameter divides a circle into two equal parts; each arc is a semicircle of 180°. If a triangle is inscribed in a semicircle, the inscribed angle is 90°, so a triangle inscribed in a semicircle is always a right triangle.

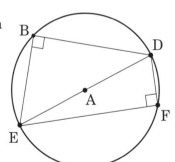

○ The opposite angles of a quadrilateral inscribed in a circle are always supplementary.

$m\widehat{OPQ} + m\widehat{ORQ} = 360°$
$\frac{1}{2}m\widehat{OPQ} + \frac{1}{2}m\widehat{ORQ} = \frac{1}{2} \cdot 360° = 180°$
$m\angle OPQ + m\angle ORQ = 180°$

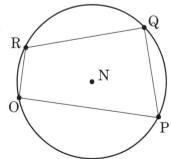

Practice

1. Find the measures of the remaining arcs and angles in this figure.

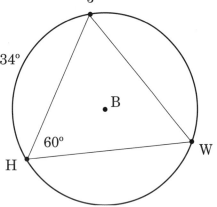

2. Find the measures of the remaining arcs and angles in this figure.

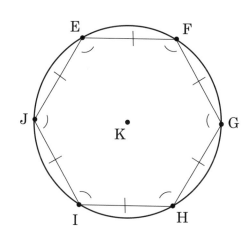

3. Find the measures of the remaining arcs and angles. *H* is the center of the circle.

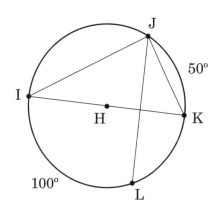

4. $\angle JIH =$
   $m\overset{\frown}{EGH} =$

5. Find the measures of all angles in $\triangle RDV$ and $\triangle ODL$.
   A is the center of the circle.

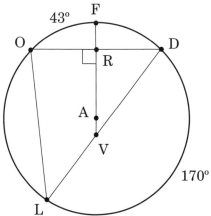

Chapter 5

Segment Lengths and Arc Measurements

Intersecting Chords

○ When two chords intersect in a circle, they form four arcs and two pairs of vertical angles. Unless the chords happen to intersect on the circle or at its center, no rule that we've learned so far will help.

○ To calculate the measure of an angle formed by two chords intersecting in a circle, first add the measures of the arcs intercepted at both ends. Then divide that sum by two.

$m\angle BEC = \frac{1}{2}\left(m\widehat{AD} + m\widehat{BC}\right) = \frac{1}{2}(70 + 100) = \frac{1}{2}(170) = 85°$.

$m\angle BEC$ and $\angle AED$ are vertical angles, so they are congruent.

$m\angle AEB = \frac{1}{2}\left(m\widehat{AB} + m\widehat{DC}\right) = \frac{1}{2}(50 + 140) = \frac{1}{2}(190) = 95°$.

$m\angle AEB$ and $\angle DEC$ are vertical angles, so they are congruent.

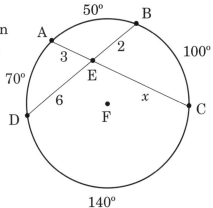

○ It is also possible to calculate the lengths of segments on these intersecting chords. The lengths can be expressed as a proportion or as an equation of products.

$\dfrac{AE}{BE} = \dfrac{ED}{EC}$ $\dfrac{3}{2} = \dfrac{6}{x}$ $AE \cdot EC = BE \cdot ED$

              $3x = 12$ or $3 \cdot x = 2 \cdot 6$

              $x = 4$ $x = 4$

Practice

1. Find the measures of all arcs and angles in $\odot F$. $m\angle AED = 111°$. Find the value of x.

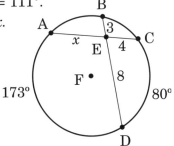

2. Find the measures of all arcs and angles in $\odot F$. Find the value of x.

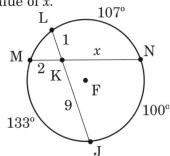

3. Find the value of x and y.

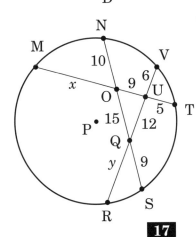

4. The radius of $\odot C$ has a length of 15mm. $\overline{LM} \cong \overline{LH}$, $\overline{JV} \cong \overline{FM}$, $m\widehat{YM} = 20°$, and $m\widehat{MH} = 75°$. Find JH, FL, $m\widehat{HV}$, and $m\angle MLH$.

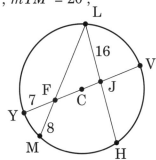

Tangents, Chords, and Arcs

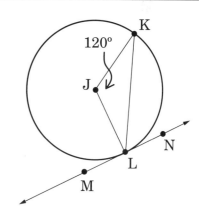

○ Just as the measure of an inscribed angle is half the measure of its intercepted arc, the measure of an angle between a chord and a tangent is half the measure of its intercepted arc.

• $\overleftrightarrow{MN}$ is tangent to $\odot J$ and $\overline{LK}$ is a chord in $\odot J$. Since $m\angle KJL = 120°$, $m\widehat{LK} = 120°$, the relationship between chords, tangents, and arcs is expressed as $m\angle KLN = \frac{1}{2}m\widehat{LK}$. Therefore $m\angle KLN = \frac{1}{2}(120) = 60°$.

In this example it is possible to verify that $m\angle KLN = 60°$. $\triangle JKL$ is an isosceles triangle (both legs are radii, and therefore congruent). Since the vertex angle is 120°, each base angle is 30°. Since the radius to the point of tangency is perpendicular to the tangent line, $\angle JLN = 90°$. $90° - 30° = 60°$, so $m\angle KLN = 60°$.

Practice

1. $m\widehat{DH} = 70°$

   $m\angle DHL =$

   $m\angle DHI =$

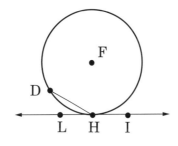

2. $m\widehat{SE} = 42°$

   $m\widehat{EK} = 88°$

   $m\angle SEQ =$

   $m\angle KEN =$

   $m\angle SEK =$

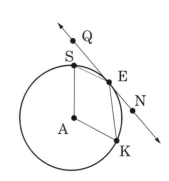

3. $m\widehat{XY} = 118°$

   $m\angle ZYX =$

   $m\angle XZY =$

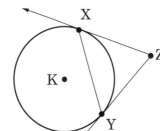

4. $m\angle MCO = 35°$

   $m\angle CPO =$

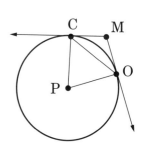

Tangents, Secants, and Arc Measurements

○ Three related equations allow us to find the lengths of intercepted arcs when two lines are drawn from a single point outside a circle. Look at the arcs and angles formed when two chords intersect in a circle. You will see similarities.

Two Secants

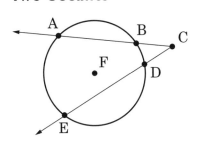

$$\tfrac{1}{2}\left(m\overset{\frown}{AE} - m\overset{\frown}{BD}\right) = m\angle C$$

Two Tangents

$$\tfrac{1}{2}\left(m\overset{\frown}{GKI} - m\overset{\frown}{GI}\right) = m\angle H$$

A Secant and a Tangent

$$\tfrac{1}{2}\left(m\overset{\frown}{LO} - m\overset{\frown}{MO}\right) = m\angle N$$

Examples:

1. $360 - (165 + 100 + 75) = 20° = m\overset{\frown}{SM}$

   $m\angle Y = \tfrac{1}{2}\left(m\overset{\frown}{VA} - m\overset{\frown}{SM}\right)$

   $\quad = \tfrac{1}{2}(100 - 20)$

   $\quad = \tfrac{1}{2}(80)$

   $\quad = 40°$

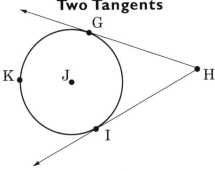

2. $360 - 115 = 245° = m\overset{\frown}{IFU}$

   $m\angle Q = \tfrac{1}{2}\left(m\overset{\frown}{IFU} - m\overset{\frown}{IU}\right)$

   $\quad = \tfrac{1}{2}(245 - 115)$

   $\quad = \tfrac{1}{2}(130)$

   $\quad = 65°$

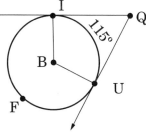

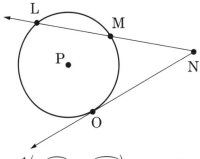

3. Take a closer look at this situation: two tangents drawn to a circle from a single point. We know that $m\angle BCA = \tfrac{1}{2}m\overset{\frown}{BC}$ and that $m\angle ABC = \tfrac{1}{2}m\overset{\frown}{BC}$. Since $m\angle BCA = \tfrac{1}{2}m\overset{\frown}{BC} = m\angle ABC$, the two angles are congruent. Thus, $\triangle BCA$ is an isosceles triangle and $\overline{AB} \cong \overline{AC}$. In fact two tangent lines can be drawn to a circle from a point outside that circle. And the segments from the point outside the circle to the points of tangency are always congruent to each other.

4. $360 - (110 + 160) = 90° = m\overset{\frown}{WE}$

   $m\angle J = \tfrac{1}{2}\left(m\overset{\frown}{WV} - m\overset{\frown}{WE}\right)$

   $\quad = \tfrac{1}{2}(160 - 90)$

   $\quad = \tfrac{1}{2}(70)$

   $\quad = 35°$

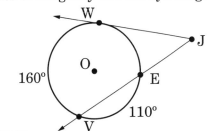

Practice

1. If $m\widehat{LVR} = 200°$, what is $m\angle Y$?

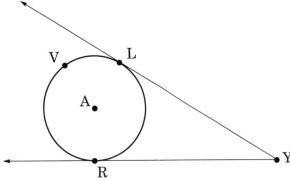

2. If $m\angle X = 53°$ and $m\widehat{WG} = 84°$, what is $m\widehat{GM}$?

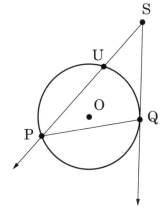

3. If $m\angle N = 38°$, $m\widehat{MJ} = 80°$, and $m\widehat{IJ} = 132°$, what is $m\widehat{IA}$?

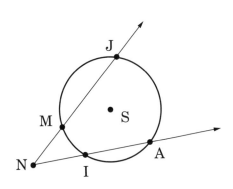

4. $m\angle USQ = 30°$

$m\widehat{PQ} = 139°$

$m\widehat{UQ} =$

$m\widehat{PU} =$

$m\angle UPQ =$

$m\angle SQP =$

5. $ED =$

$EG =$

$EJ =$

$m\widehat{JD} =$

$m\angle G =$

$m\angle JED =$

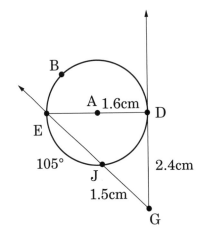

T e s t i n g   f o r   T a n g e n c y

○ Tangent lines intersect a circle in exactly one point. Up until now, we have simply decided that lines were tangent by looking at them, or by noting that they were perpendicular to the radius at the point of tangency. But it is possible to mathematically verify that a line is tangent to a circle.

• If $\overleftrightarrow{BC}$ is a tangent, then it must be perpendicular to $\overline{AB}$.
If they are perpendicular, then $\triangle ABC$ is a right triangle.
Therefore, to test whether or not $\overleftrightarrow{BC}$ is a tangent, use the converse of the Pythagorean Theorem: if $AB^2 + BC^2 = AC^2$, then this is a right triangle and $\overleftrightarrow{BC}$ is a tangent.

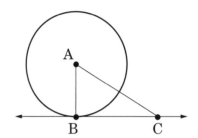

Example:
$$8^2 + 15^2 = 17^2$$
$$64 + 225 = 289$$
$$289 = 289$$
Therefore, $\overleftrightarrow{FQ}$ is a tangent.

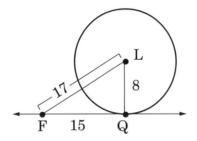

Practice

Determine whether or not each line is a tangent.

1.

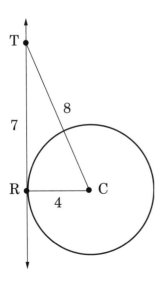

2.

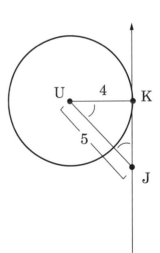

3.

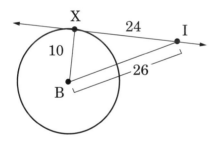

4.

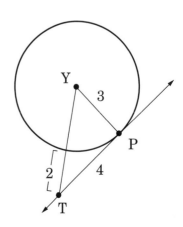

21

T a n g e n t , S e c a n t ,
a n d S e g m e n t L e n g t h s

○ Just as there are ways to find the measures of angles and arcs formed by tangents and secants, there are also ways to calculate the lengths of segments. These calculations are very similar to the formula for finding the lengths of segments on intersecting chords.

Two Secants

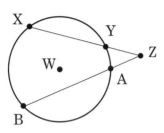

$$XZ \cdot YZ = BZ \cdot AZ$$

Secant and Tangent

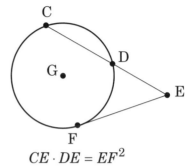

$$CE \cdot DE = EF^2$$

○ When two secants begin at the same point, they form four segments: two outside the circle and two inside the circle. Their relationship can be expressed as an equation, as shown above, or as a proportion: $\frac{secant_1}{secant_2} = \frac{extension_2}{extension_1}$ or $\frac{XZ}{BZ} = \frac{AZ}{YZ}$.

○ The relationship for a secant and a tangent is closely related. But the entire tangent segment is outside the circle, so that side of the equation is simply the length of the tangent segment squared. (In fact, the length of the tangent segment is the geometric mean of the entire secant segment and the exterior portion of the secant segment.)

Examples:

1. Since $\overline{KJ}$ is a tangent segment and $\overline{KI}$ is a secant segment, $KH \cdot KI = KJ^2$
$$4 \cdot 16 = KJ^2$$
$$64 = KJ^2$$
$$KJ = 8$$

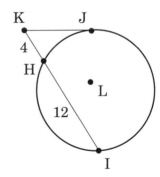

2. Since $\overline{ER}$ and $\overline{EN}$ are secant segments,
$$EV \cdot ER = EA \cdot EN$$
$$2 \cdot 12 = x \cdot 8$$
$$24 = 8x$$
$$x = 3$$

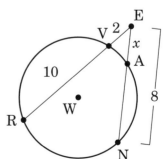

Practice

1. Find the value of *x*.

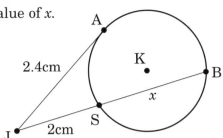

2. Find the value of *x*.

$m \angle B =$

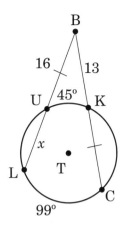

3. $\overline{MD}$ is tangent to $\odot P$.
Find the value of *x*.
Find the measures of the
angles in the triangle.

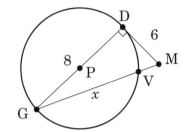

4. Find the value of *x*.

$m \angle FOP =$
$m\widehat{YX} =$
$m\widehat{YF} =$
$m\widehat{FG} =$
$m\widehat{GX} =$

5. The radius has a length of 13.
$\overline{RP}$ and $\overline{HP}$ are tangent to $\odot Y$.

$m \angle RPH =$
$m\widehat{HR} =$
$m\widehat{KR} =$
$ZP =$
$x =$
$ZR =$

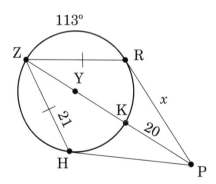

6. Find the lengths of the other segments.

$CS = 2.5$
$LM = 5.0$
$LA = 1.6$
$AW = 3.8$
$LQ = 1.7$

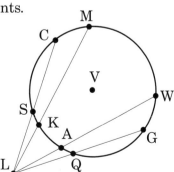

7. $\overline{GP}$ and $\overline{HY}$ are tangents to $\odot K$.

$HK = 14$

$ZG = 18$

$GP = 26$

$m\widehat{ZP} = 88°$

$HG =$

$m\angle G =$

$m\angle HPY =$

$HY =$

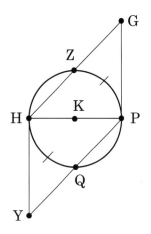

8. $\overline{BK}$ is tangent to $\odot T$.

$BS = 5$

$KT = 13$

$BK = 11$

$NW = 14$

$m\widehat{SK} = 51°$

$m\widehat{SE} = 109°$

$m\widehat{EN} =$

$m\angle W =$

$m\angle B =$

$BW =$

$KW =$

$EW =$

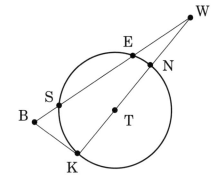

Chapter 6

Tangent Circles and Common Tangents

Tangent Circles

○ Circles can be internally or externally tangent.

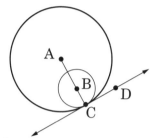

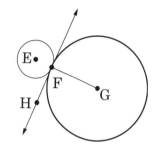

⊙A and ⊙B are **internally tangent** because
- they intersect in exactly one point
- they are both tangent to the same line
- one circle is entirely contained in the other

⊙E and ⊙G are **externally tangent** because
- they intersect in exactly one point
- they are both tangent to the same line
- they do not share any internal space

○ The centers of tangent circles lie on a line. Since tangent circles are tangent to a single line, their radii must be perpendicular to the same line. If they are to have radii perpendicular to the same line and either completely shared internal space or have completely distinct internal space, their centers must lie on a line.

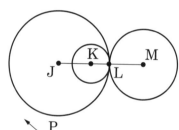

Practice

1. Find the value of x.

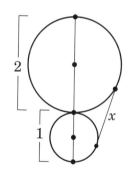

2. Name all pairs of tangent circles.

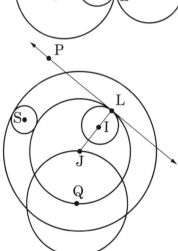

3. Find the length of all segments.

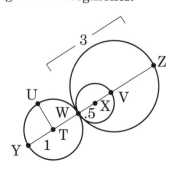

4. Find the value of x and y.

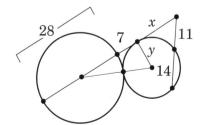

Common Tangents

○ A line can be tangent to more than one circle at the same time. This can take two forms:

An internal tangent passes between the circles. If you imagine a segment connecting the centers of the circles, an internal tangent will intersect that segment.

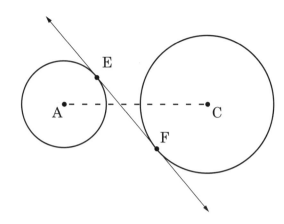

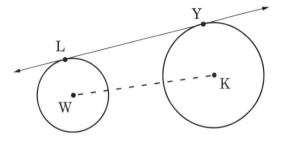

An external tangent appears to be sitting on the circles, or the circles appear to be sitting on the tangent line. If you imagine a segment connecting the centers of the circles, an external tangent will not intersect that segment.

Practice

I. If *EI = 5* and *IF = 3*, find the lengths of $\overline{IH}$, $\overline{EH}$, $\overline{GI}$ and $\overline{FG}$.

2. Name any congruent segments. Draw a common internal tangent.

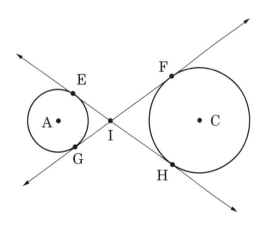

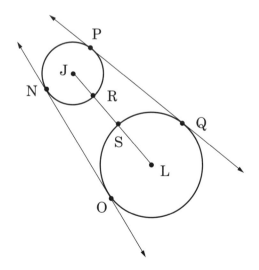

3. What types of common tangents are in this picture? Name the pairs of common tangents. If $BD = 4$, are there other segments whose lengths you can find? Why or why not?

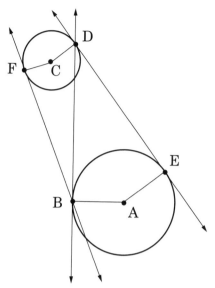

4. If $GD = 1.8$cm, $EF = 1.6$cm, and $AG = 5$cm, find $AF = $ ____,

$DF = $ ____,

$m \angle AFD = $ ____.

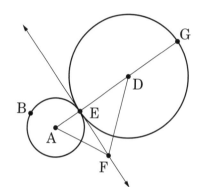

5. Draw two circles (use a compass or trace something circular) and name them A and B. Draw and label a common internal and a common external tangent. On $\odot A$ draw a tangent which is not a common tangent for the two circles. Draw a right triangle by drawing a radius of $\odot A$ and then connecting the center of the circle with a point on the tangent line. Use a ruler to measure the lengths of the sides of that right triangle. Use the Pythagorean Theorem to test your tangent line to see if it really is a tangent.

Chapter 7

Area and Circumference

Circumference and Area Defined

○ *Circumference* is the distance around a circle. The formula for calculating the circumference of a circle is $C = 2\pi r$ or $C = \pi d$. C stands for circumference; r stands for radius; d stands for diameter; π can be approximated with 3.14, but it can also be left as a variable.

○ *Area* is the number of square units it would take to cover the surface of the circle. The formula for calculating the area of a circle is $A = \pi r^2$. A stands for area.

If the radius (r) of this circle is 12 inches,

$$C = 2\pi r \qquad \text{and} \qquad A = \pi r^2$$
$$= 2\pi 12 \text{ in} \qquad \qquad = \pi 12^2$$
$$= 24\pi \text{ in} \qquad \qquad = 144\pi \text{ in}^2$$

Practice

Calculate the area and perimeter of each circle.

1.

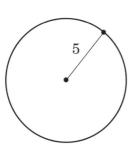

3 in

2.

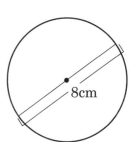

8cm

3.

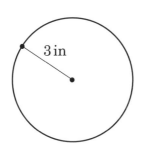

5

4.

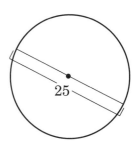

25

5.

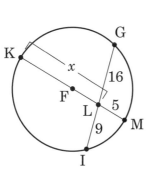

6.

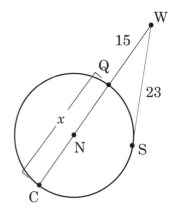

A r e a s   a n d   I n s c r i b e d   P o l y g o n s

○ By applying the area formulas for circles and polygons, we can find the area of shapes involving circles, parts of circles, and areas around inscribed figures.

Examples:

To find the area of the shaded region,

1. Find the area of the triangle.
2. Calculate the length of the diameter.
3. Find the area of the circle.
4. Subtract the area of the triangle from the area of the circle.

•Figure 1

1. Since this is a right triangle, the legs are perpendicular and therefore serve as the base and height.
2. Using the Pythagorean Theorem, the diameter is 17. Therefore the radius is $\frac{17}{2}$ = 8.5.
3. $A = \pi r^2$
    $= \pi (8.5)^2$
    $= 72.25\pi$
4. Area = $72.25\pi - 60 \approx 167$.

$A = \frac{1}{2}bh$
$ = \frac{1}{2}(8)(15)$
$ = 60$

$a^2 + b^2 = c^2$
$15^2 + 8^2 = VW^2$
$225 + 64 = VW^2$
$289 = VW^2$
$17 = VW$

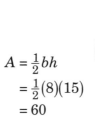

Figure 1

To find the area of the shaded region,

1. Find the area of the circle.
2. Use the formula for finding the area of a square after using 45–45–90 triangles to find the length of a side.
3. Subtract the area of the square from the area of the circle.

•Figure 2

1. $A = \pi r^2 = \pi 3^2 = 9\pi$
2. Draw a second radius to form a 45–45–90 triangle in the square. Since each radius is a leg and the side of the square is the hypotenuse, use the fact that the hypotenuse = leg $\cdot \sqrt{2}$. Thus, the length of a side of the square is $3\sqrt{2}$. And the area is $\left(3\sqrt{2}\right)^2 = 9 \cdot 2 = 18$.
3. Area = $9\pi - 18 \approx 10.3$.

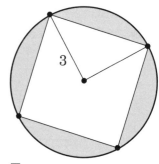

Figure 2

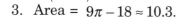

Practice

Find the area of the shaded regions

1.

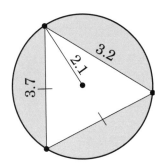

2.

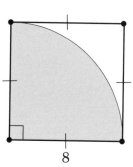

3.

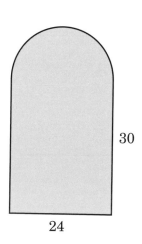

4.

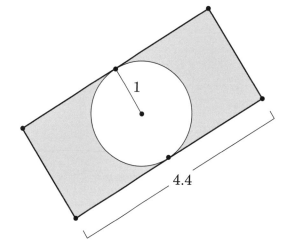

5.

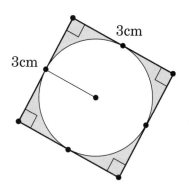

6. $\triangle PNO$ is an equilateral triangle.
The diameter of the large circle is 16.

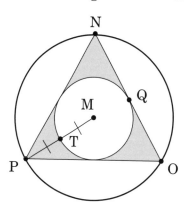

31

Chapter 8

Sector Area and Arc Length

Sector of a Circle

- The formula, remember, for the area of a circle is $A = \pi r^2$.
- A central angle cuts off a section of the circle which is bounded by the rays of the central angle and their intercepted arc. When a central angle defines a region of a circle, it is called a **sector of a circle**. To get a mental picture of a sector, think of a piece of pie.
- A sector is a portion of the circle. A central angle contains a portion of the 360° that make up a circle. Therefore the formula for the area of a sector is based on finding the portion of the area of the larger circle based on the central angle. If N represents the measure of the central angle: $A = \frac{N}{360} \pi r^2$.

Examples:

1. If $r = 4$ and $N = 30°$

   $A = \frac{30}{360} \pi 4^2$

   $= \frac{1}{12} \pi 16$

   $= \frac{16}{12} \pi$

   $= \frac{4}{3} \pi$

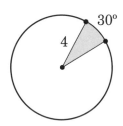

2. The measure of the minor arc is $360° - 200° = 160°$.
   The measure of the central angle equals that of the minor arc.

   If $r = 7$ and $N = 160°$

   $A = \frac{160}{360} \pi 7^2$

   $= \frac{4}{9} \pi 49$

   $= \frac{196}{9} \pi$

   $\approx 21.8\pi$

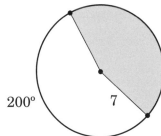

Practice:

Find the area of the shaded region.

1.

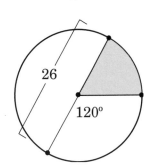

2.

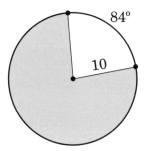

3.

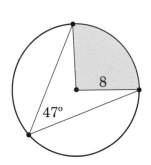

4.

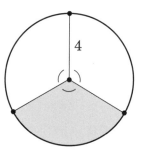

5.

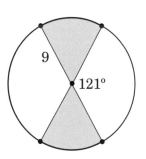

6.

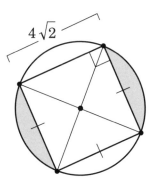

Arc Length

○ The formula, remember, for the circumference of a circle is $C = 2\pi r$.

○ An arc is a part of a circle. Often arc measurements are given in terms of degrees, corresponding to their central angles.

○ Sometimes we actually want to know the distance around a circle in feet or miles. Like, how far would we travel around the earth, or how much wood would it take to build an arch? In such a case, we need to calculate the **arc length**.

○ Just as a sector comprises a portion of the arc of a circle, arc length comprises a portion of the circumference.

Examples:

1. For this circle with radius $XY = 10$,
   the arc length of $\overset{\frown}{XZ} = \frac{m\overset{\frown}{XZ}}{360} \cdot 2\pi r$,
   so the arc length of $\overset{\frown}{XZ} = \frac{40}{360} \cdot 2\pi 10$
   $$= \frac{20}{9}\pi$$
   $$\approx 2.2\pi$$

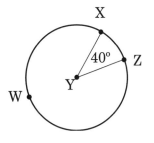

2. To find the arc length of $\overset{\frown}{RT}$,
   $m\overset{\frown}{RUT} = 246°$,
   so $m\overset{\frown}{RT} = 360° - 246° = 114°$.
   Therefore the measure of $\angle RST = 114°$.
   Use the formula for arc length:
   arc length of $\overset{\frown}{RT} = \frac{114}{360} \cdot 2\pi 12$
   $$= \frac{57}{180} \cdot 24\pi$$
   $$= \frac{1368}{180}\pi$$
   $$= 7.6\pi$$

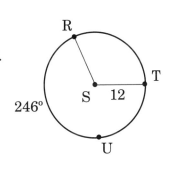

Practice

Find the length of the arcs or measures of the angles (x and / or y) in each figure.

1.

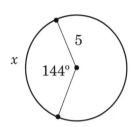

2.

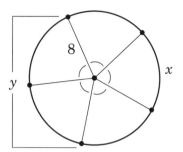

3.

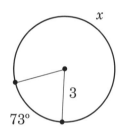

4.

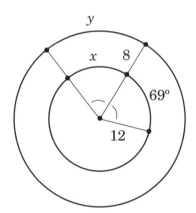

5.

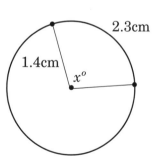

6.

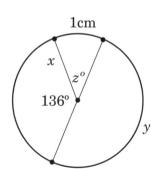

Chapter 9

Equations and Graphing

The Equation of a Circle

○ **The standard equation for a circle is $(x - h)^2 + (y - k)^2 = r^2$.**
The center of the circle is the point (h, k).
The length of the radius is r.

○ **Writing an equation for a circle:**
- If the center of the circle is $(3, 5)$ and the radius is 2, then $h = 3$, $k = 5$ and $r = 2$. Substituting into the standard equation, $(x - 3)^2 + (y - 5)^2 = 2^2$, the simplified equation is $(x - 3)^2 + (y - 5)^2 = 4$.
- If the center of the circle is $(-2, 4)$ and the radius is 8, then $h = -2$, $k = 4$ and $r = 8$.
Substituting, $(x - (-2))^2 + (y - 4)^2 = 8^2$, and $(x + 2)^2 + (y - 4)^2 = 64$.

○ **Writing an equation for a circle given the center and a point on the circle:**
- If the center of a circle is $(4, 3)$ and the point $(6, -1)$ is a point on the circle, we first need the **distance formula** to calculate the radius of the circle.
$$r = \sqrt{(6 - 4)^2 + (-1 - 3)^2} = \sqrt{2^2 + (-4)^2} = \sqrt{4 + 16} = \sqrt{20} = 2\sqrt{5}$$
Then substitute into the standard equation of a circle: $(x - 4)^2 + (y - 3)^2 = \left(2\sqrt{5}\right)^2$.
Simplifying, $(x - 4)^2 + (y - 3)^2 = 20$.

Practice

Write an equation for each circle based on the given information.

1. Center $(-3, 0)$, radius 3.

2. Center $(0, 0)$, radius 4.

3. Center $(8, -4)$, radius 1.

4. Center $(1, 2)$, point on the circle $(4, 4)$.

5. Center $(-1, -7)$, point on the circle $(2, -2)$.

6. Center $(2, -3)$, point on the circle $(-4, 5)$.

Graphing Circles

○ The graph of a circle is **based on the center and radius** of the circle.

To graph a circle with a center of (–3, 1) and a radius of 4:
- graph the point (–3, 1)
- put the point of your compass on (–3, 1) and set its width to 4 units. The pencil should reach (1, 1) because that point is 4 units to the right of (–3, 1).
- draw the circle

To graph a circle with a center of (2, 2) and a radius of 5.
- graph the point (2, 2)
- set your compass so that it has a width of 5 units —options for points to line it up include (7, 2), (–3, 2), (2, 7) and (2, –3) because those are the points that are 5 units to the left, right, above and below the center.

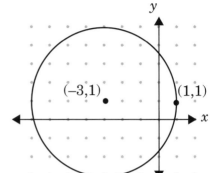

○ To graph a circle **based on an equation**, you must identify the center and radius. For a circle with the equation $(x - 2)^2 + (y - 3)^2 = 9$, the center is (2, 3) and the radius is found by solving $r^2 = 9$, so $r = 3$.

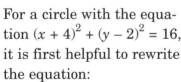

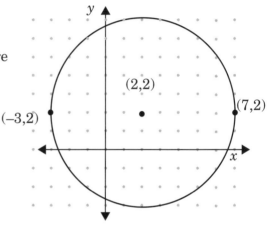

For a circle with the equation $(x + 4)^2 + (y - 2)^2 = 16$, it is first helpful to rewrite the equation:
$(x - (-4))^2 + (y - 2)^2 = 4^2$
so that the radius is 4 and the center is (–4, 2).

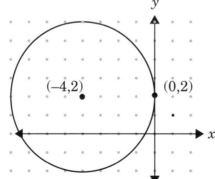

For the circle with the equation of $(x + 3)^2 + y^2 = 28$, rewrite the equation $(x - (-3))^2 + (y - 0)^2 = (2\sqrt{7})^2$ so the center is (-3, 0) and the radius is $2\sqrt{7}$ which is approximately 5.3.

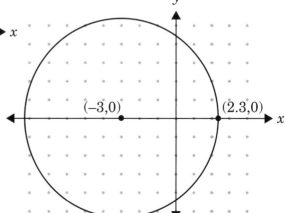

Practice

Graph these circles on the grids provided (4 circles per grid).

1. Center $(0, 0)$, radius 1.

2. Center $(2, 3)$, radius 2.

3. Center $(-3, 3)$, radius 2.

4. Center $(0, -2)$, diameter 6.

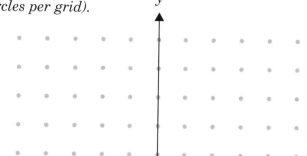
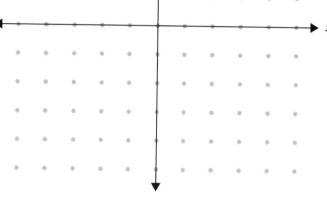

Each dot represents one unit.

5. $(x - 2)^2 + (y - 8)^2 = 25$

6. $(x - 1)^2 + (y + 4)^2 = 4$

7. $(x + 5)^2 + y^2 = 49$

8. $x^2 + y^2 = 1$ (After you graph it, explain why this circle is called the unit circle.)

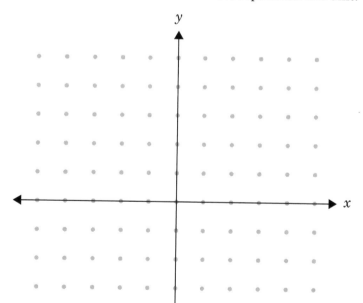

Each dot represents two units.

39

Circles Exam

Given that $\overline{DA} \perp \overline{AB}$ ...

1. Name a chord that is not a diameter.

2. Name a diameter.

3. Name a radius.

4. Name a tangent.

5. Name a secant.

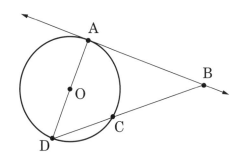

Given that $\overline{CA}$ and $\overline{CB}$ are tangent to $\odot O$ ...

6. $m \angle ADB =$

7. $m \angle AOB =$

8. $m \angle ACB =$

9. $m \angle EAO =$

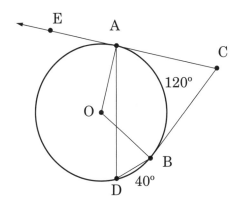

10. If $m \angle AEB = 20°$, then $m \angle AFB =$

11. If $m\overarc{AB} = 95°$ and $m\overarc{EF} = 25°$, then $m \angle AGB =$

12. If $m\overarc{AB} = 70°$ and $m\overarc{EF} = 30°$, then $m \angle AHB =$

13. If $m\overarc{AB} = 85°$ and $m \angle EHF = 59°$, then $m\overarc{EF} =$

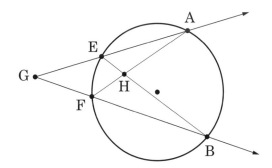

14. Find x.

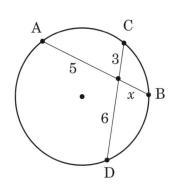

15. Find x.

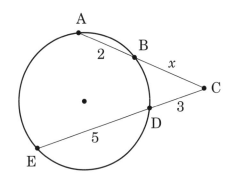

16. $\overline{BA}$ is tangent to the circle. Find x.

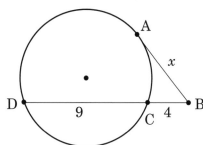

17. $\overline{CA}$ and $\overline{CB}$ are tangent to the circle. Find x.

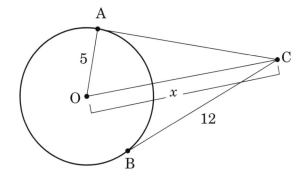

18. Find the area of a circle with a diameter of 18.

19. Find the area of a circle whose circumference is $22\pi\,cm$.

20. Find the area of sector ACB of $\odot C$ if the radius is 8 and $m\overset{\frown}{AB} = 40°$.
    You may want to draw your own picture before doing any calculations.

21. Find the area of the shaded region.

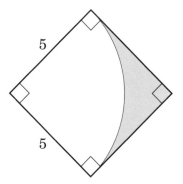

22. In $\odot O$, $m\overset{\frown}{AB} = 72°$ and the area of sector AOB is 20π. Find the length of $\overset{\frown}{AB}$.

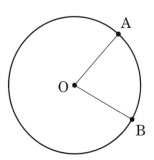

Glossary

Auxiliary Line: A line that is added to a picture. It is sometimes helpful to extend a line or add the ray of an angle to solve a geometric problem. Auxiliary lines are usually dashed so that it is clear that they aren't part of the original problem.

Complementary Angles: Two angles whose sum is 90°.

 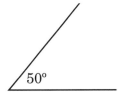 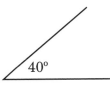

Converse Statement: A new statement formed by reversing an original statement's hypothesis and conclusion. Note: a converse statement is not necessarily true. For example: Statement-If a quadrilateral is a square, then it has four right angles (True). Converse statement-If a quadrilateral has four right angles, then it is a square (False. It could be a rectangle).

Distance Formula: To find the distance (d) between two points, (x_1, y_1) and (x_2, y_2) substitute the points into the formula $d = \sqrt{(x_2 - y_1)^2 + (y_2 - x_1)^2}$ and simplify.

Parallel Lines: Two lines which do not intersect.
○ $\overleftrightarrow{AC} \| \overleftrightarrow{DF}$ is read as "line AC is parallel to line DF."
○ $\overleftrightarrow{GH}$ is a transversal because it intersects 2 lines.
○ The arrows on the lines show that they are parallel.
○ When 2 parallel lines are cut by a transversal:

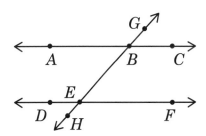

    a. alternate interior angles are congruent
       $\angle CBE \cong \angle BED$ and $\angle ABE \cong \angle BEF$.
    b. corresponding angles are congruent
       $\angle GBA \cong \angle BED$ and $\angle GBC \cong \angle BEF$
       and $\angle ABE \cong \angle DEH$ and $\angle CBE \cong \angle FEH$.
    c. same-side interior angles are supplementary
       $m\angle ABE + m\angle DEB = 180°$ and $m\angle CBE + m\angle FEB = 180°$.

Parallelogram: A quadrilateral in which both pairs of opposite sides are parallel.
○ Pairs of opposite sides are congruent.
○ Pairs of opposite angles are congruent.
○ Diagonals bisect each other.

Rectangle: A parallelogram with 4 right angles.
○ All of the properties of parallelograms also apply to rectangles.
○ Diagonals are congruent to each other.

Square: A rectangle with 4 congruent sides.
○ All properties of parallelograms and rectangles apply to squares.
○ Diagonals intersect to form right angles.
○ Diagonals bisect the angles of the parallelogram.

Supplementary Angles: Two angles whose sum is 180º.

○ If two adjacent angles form a straight line, they are supplementary.

 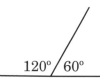

Pythagorean Theorem: In a right triangle, if the lengths of two sides are known, then the length of the third side can be found. There are three variables (a, b, and c): c always represents the hypotenuse. If you know any two of the variables, you can use them in the formula to calculate the third variable: $a^2 + b^2 = c^2$.

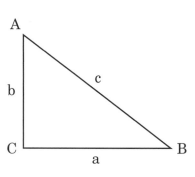

Quadratic Formula: This formula is used to find the solutions to quadratic equations when factoring is not possible. a = coefficient of the 2nd degree term, b = the coefficient of the 1st degree term, c = the constant term, and x = the variable in the equation.

$$x = \frac{-b \pm \sqrt{b^2 - 4ac}}{2a}$$

Special Right Triangles:

45–45–90 Triangles	30–60–90 Triangles
(Isosceles Right Triangles)	

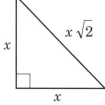

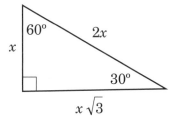

If you know one leg of a 45–45–90 triangle, you automatically know the other leg because they are congruent to each other. But you can find the hypotenuse right away because it will always be $\sqrt{2}$ times the length of the leg.

If you know the length of the shortest leg in a 30–60–90 triangle, you can find the longer leg by multiplying the length of the shorter leg by $\sqrt{3}$. You can find the length of the hypotenuse by doubling the shorter leg.

Trigonometry: The lengths of the legs of right triangles are in special relationships to the measures of the acute angles of the triangle. These relationships are expressed as the tangent, sine and cosine functions. **SohCahToa** is a useful trick for remembering the 3 functions: **S**ine is **o**pposite over **h**ypotenuse, **C**osine is **a**djacent over **h**ypotenuse, and **T**angent is **o**pposite over **a**djacent. Look at the picture and the ratios below.

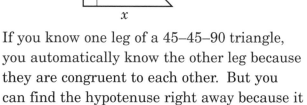

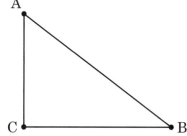

If you need to find the measure of an angle in a right triangle and you know the measures of the sides, you will either need a scientific calculator or a trigonometric values table. On a calculator you will type something like $\tan^{-1}.976 =$ (if .976 is what you got when you took the ratio of the legs) and the calculator will give you the angle measure. Otherwise you will look the decimal up in the table and read off your angle.

= equals   ≅ congruent   ~ similar   ≈ approximately π pi

Answer Key

Page 4

	1.	2.	3.
Circle	$\odot C$	$\odot S$	$\odot B$
Chord	$\overline{JV}$, $\overline{WD}$, $\overline{VD}$	$\overline{EA}$, $\overline{FK}$	$\overline{GH}$, $\overline{MN}$, $\overline{GI}$
Diameter	$\overline{JV}$, $\overline{WD}$	$\overline{FK}$	$\overline{MN}$
Radius	$\overline{WC}$, $\overline{JC}$, $\overline{CV}$, $\overline{CD}$	$\overline{MS}$, $\overline{ST}$, $\overline{SK}$	$\overline{DB}$, $\overline{BM}$, $\overline{BN}$
Right Angle	$\angle MDC$, $\angle TDC$	$\angle MST$, $\angle STB$, $\angle BTR$, $\angle RTJ$, $\angle JTS$	none
Secant	$\overline{JV}$, $\overline{WD}$, $\overline{VD}$	$\overleftrightarrow{FK}$, $\overline{EA}$	$\overleftrightarrow{MN}$, $\overline{GH}$, $\overline{GI}$
Tangent Point of Tangency	$\overleftrightarrow{MT}$ D	$\overleftrightarrow{BJ}$ T	none none

	3.	4.	5.
Circle	$\odot D$	$\odot L$	$\odot W$
Chord	$\overline{CK}$, $\overline{CE}$	$\overline{KE}$	$\overline{LV}$, $\overline{ZD}$
Diameter	$\overline{CE}$	none	$\overline{LV}$
Radius	$\overline{DE}$, $\overline{CD}$	$\overline{KL}$, $\overline{LC}$, $\overline{LE}$	$\overline{WL}$, $\overline{WC}$, $\overline{WV}$
Right Angle	none	none	none
Secant	$\overline{CE}$, $\overline{CK}$	$\overline{KE}$	$\overleftrightarrow{LV}$, $\overleftrightarrow{ZD}$
Tangent Point of Tangency	$\overline{GH}$ E	$\overline{GD}$, $\overline{DF}$, $\overline{GF}$ K, C, E	$\overleftrightarrow{QB}$ C

	5.	6.	6.
Circle	$\odot A$	$\odot M_1$	$\odot M_2$
Chord	$\overline{MP}$, $\overline{NC}$, $\overline{OR}$	$\overline{BT}$	$\overline{KS}$, $\overline{LS}$, $\overline{RC}$
Diameter	$\overline{NC}$	$\overline{BT}$	$\overline{RC}$
Radius	$\overline{AN}$, $\overline{AC}$	$\overline{BM}$, $\overline{MW}$, $\overline{MJ}$, $\overline{MT}$	$\overline{MR}$, $\overline{MC}$, $\overline{MG}$
Right Angle	none	none	none
Secant	$\overleftrightarrow{ZD}$, $\overline{NC}$, $\overline{OR}$	$\overline{BT}$	$\overleftrightarrow{KS}$, $\overleftrightarrow{LS}$, $\overline{RC}$
Tangent Point of Tangency	$\overleftrightarrow{QB}$ C	$\overleftrightarrow{KS}$, $\overleftrightarrow{LS}$ T, W	none none

Page 5

1. All radii are congruent in a circle, so $ON = 15$ and $NM = 15$. Since a diameter is twice as long as a radius, $OM = 30$.

2. $AD = 3$, $AC = 3$, $AB = 1.5$, $BC = 1.5$.

3. $\angle VZX$ is a right angle because $\overleftrightarrow{ZX}$ is a tangent, so $\triangle VZX$ is a right triangle. $VZ = 8$, because $VW = 8$. By the Pythagorean Theorem, $8^2 + 3^2 = VX^2$, so $64 + 9 = VX^2$, $73 = VX^2$, and $VX \approx 8.5$. $VX - VY = YX$, so $YX \approx 8.5 - 8 \approx .5$.

4. $EH = 8$. $EF = FG + GE = 9 + 8 = 17$. $EF^2 = FH^2 + EH^2 = 17^2 = FH^2 + 8^2 = 289 = FH^2 + 64$. So, $FH^2 = 225$ and $FH = 15$. $FH - FK = HK = 15 - 9 = 6$.

Page 7

1. $\odot Q$

minor arcs	$\overarc{ST}$, $\overarc{RS}$, $\overarc{RT}$, $\overarc{SU}$, $\overarc{UR}$, $\overarc{TU}$
major arcs	$\overarc{SRU}$, $\overarc{SRT}$, $\overarc{UTR}$, $\overarc{TSU}$, $\overarc{RUT}$, $\overarc{RUS}$
semicircles	none
central angles	$\angle SQT$, $\angle SQU$, $\angle TQU$

2. $\odot D$

minor arcs	$\overarc{AE}$, $\overarc{EB}$, $\overarc{BF}$, $\overarc{BC}$, $\overarc{AB}$, $\overarc{CF}$, $\overarc{CA}$, $\overarc{CE}$, $\overarc{FA}$
major arcs	$\overarc{FEB}$, $\overarc{BCE}$, $\overarc{EBC}$, $\overarc{EBA}$, $\overarc{ABC}$, $\overarc{ABF}$, $\overarc{BFA}$, $\overarc{CAB}$, $\overarc{CAF}$
semicircles	$\overarc{FEB}$, $\overarc{FCE}$, $\overarc{FAE}$
central angles	$\angle FDA$, $\angle ADE$, $\angle FDE$

3. $\odot V$

minor arcs	$\overarc{AB}$, $\overarc{BK}$, $\overarc{KR}$, $\overarc{AR}$, $\overarc{RB}$, $\overarc{AK}$
major arcs	$\overarc{AKR}$, $\overarc{BKR}$, $\overarc{BKA}$, $\overarc{KRA}$, $\overarc{KRB}$, $\overarc{RAK}$
semicircles	none
central angles	$\angle AVR$

4. $\odot A$

minor arcs	$\overarc{IB}$, $\overarc{EJ}$, $\overarc{EB}$, $\overarc{JB}$, $\overarc{JD}$, $\overarc{BD}$, $\overarc{DK}$, $\overarc{BK}$, $\overarc{DI}$, $\overarc{KI}$, $\overarc{KE}$, $\overarc{IE}$, $\overarc{IJ}$, $\overarc{JK}$
major arcs	$\overarc{JDI}$, $\overarc{BDI}$, $\overarc{DKJ}$, $\overarc{DKB}$, $\overarc{KIJ}$, $\overarc{KIB}$, $\overarc{KID}$, $\overarc{IEK}$, $\overarc{EBK}$, $\overarc{IJD}$
semicircles	$\overarc{EJD}$, $\overarc{EID}$
central angles	$\angle BAJ$, $\angle BAE$, $\angle BAD$, $\angle JAE$, $\angle JAD$, $\angle DAE$

4. $\odot E$

minor arcs	$\overarc{JG}$, $\overarc{GI}$, $\overarc{JI}$, $\overarc{IH}$, $\overarc{IC}$, $\overarc{HC}$, $\overarc{HJ}$, $\overarc{CJ}$, $\overarc{GH}$
major arcs	$\overarc{GIJ}$, $\overarc{IHJ}$, $\overarc{IHG}$, $\overarc{HCG}$, $\overarc{HCI}$, $\overarc{JIC}$, $\overarc{CJI}$, $\overarc{CJH}$, $\overarc{JGH}$
semicircles	$\overarc{GJC}$, $\overarc{GIC}$, $\overarc{GHC}$
central angles	$\angle IEH$, $\angle IEC$, $\angle CEH$, $\angle GEI$, $\angle GEH$, $\angle CEG$

Page 10

1. $m\overarc{AC} = 64°$. $m\overarc{ADC} = 360° - 64° = 296°$.

2. $m\overarc{YL} = 50°$. $m\overarc{YLM} = 180°$. $m\angle LXM = 180° - 50° = 130°$. $m\angle YXM = 180°$. $m\angle ZXY = 87°$. $m\angle ZXL = 87° + 50° = 137°$.

3. $m\angle ANV = 137°$. $m\angle AND = 180° - 137° = 43°$. $m\overarc{AD} = 43°$. $m\overarc{DQV} = 180°$.

4. $m\overarc{FG} = 25°$. $m\overarc{EG} - m\overarc{FG} = 75 - 25 = 50$, so $m\angle EJF = 50°$. $m\overarc{FH} - m\overarc{FG} = 65° - 25 = 40$, so $m\overarc{GH} = 40° = m\angle GJH$.

5. *adjacent arcs:* $\overarc{ZP}$ is adjacent to $\overarc{PN}$ and $\overarc{UZ}$; $\overarc{UN}$ is adjacent to $\overarc{UZ}$ and $\overarc{NP}$; $\overarc{NZ}$ is adjacent to $\overarc{NP}$ and $\overarc{ZP}$; $\overarc{UP}$ is adjacent to $\overarc{UN}$ and $\overarc{PN}$.

nonadjacent arcs: $\overarc{UZ}$ is not adjacent to $\overarc{NP}$ and $\overarc{ZP}$ is not adjacent to $\overarc{UN}$.

major arcs: $\overarc{ZNP}$, $\overarc{UNP}$, $\overarc{NPZ}$, $\overarc{NPU}$, $\overarc{PZN}$.

minor arcs: $\overarc{UZ}$, $\overarc{UP}$, $\overarc{ZP}$, $\overarc{PN}$, $\overarc{NU}$, $\overarc{NZ}$.

acute angles: $\angle UGZ$, $\angle ZGP$, $\angle NGU$.

obtuse angles: $\angle UGP$, $\angle PGN$, $\angle NGZ$.

6. $m\angle TRL = 35°$, because if $m\overarc{CT} = 35°$, then $m\angle CRT = 35°$ and $m\angle TRL = m\angle CRT$. $m\overarc{CL} = 2(m\overarc{CT}) = 2 \cdot 35 = 70°$. $m\angle CRH = m\angle LRB = 80° = m\overarc{LB}$. $m\angle CRB = m\angle CRT + m\angle TRL + m\angle LRB = 35 + 35 + 80 = 150°$. $m\overarc{HB} = 360 - (80 + 35 + 35 + 80) = 130°$. $m\overarc{CHB} = 80° + 130° = 210°$. $m\angle HRB = 130°$.

Page 12

1. $FD = 22$. $\overline{BC} \cong \overline{CE} \cong \overline{DF}$. $m\overarc{CE} = m\overarc{BC} = m\overarc{DF}$.

2. $UP = 25$, so $SU = UX = 25$; $SQ = 6$, so $QU = SU - SQ = 25 - 6 = 19$. Since $ZU = 19$ and $QU = 19$, $\overline{WY}$ and $\overline{RT}$ are equidistant from the center and are congruent. Since $\overline{US}\perp\overline{RT}$ and $\overline{WY} \perp \overline{UX}$, both chords are bisected by the radii and $\overline{RQ} \cong \overline{QT} \cong \overline{WZ} \cong \overline{ZY}$. Taking into account the congruent arcs, $\overarc{SR} \cong \overarc{ST} \cong \overarc{WX} \cong \overarc{XY}$ and $\overarc{RT} \cong \overarc{WY}$.

3. $m\overarc{KJ} = m\overarc{JI} = 34°$.
$m\overarc{KI} = m\overarc{JI} + m\overarc{JK} = 34° + 34° = 68°$.
$m\overarc{KLI} = 360° - 68° = 292°$. There aren't any congruent chords, because $\overline{LN}$ is not perpendicular to $\overline{XO}$, so the distance from the center cannot be determined.

4. Since $\overline{IT} \cong \overline{KR}$, $\overarc{IT} \cong \overarc{KR}$. $\overline{BN} \cong \overline{NA}$ because $\overline{QU}\perp\overline{BA}$, therefore $m\overarc{BU} \cong m\overarc{UA}$. If $ZU = 6$, then $QU = 3$. $QU - NU = QN = 3 - 1 = 2$cm. Therefore, $\overline{AB}$ is 2cm from the center of the circle.

Pages 13–14

1. If $m\overarc{BDC} = 206°$, then $m\overarc{BC} = 360° - 206°$. $m\angle BDC = \frac{1}{2}m\overarc{BC} = \frac{1}{2}(154°) = 77°$.

2. $m\angle NWT = \frac{1}{2}\overarc{NT} = \frac{1}{2}(42°) = 21°$.

3. $m\overarc{PX} + m\overarc{XE} = m\overarc{PE} = 91° + 47° = 138°$. $m\angle PBE = \frac{1}{2}m\overarc{PE} = \frac{1}{2}(138°) = 69°$.

4. $m\overarc{RC} = 81° + 58° = 139°$. $m\overarc{RQC} = 360° - m\overarc{RC} = 360° - 139° = 221°$. $m\angle RGC = \frac{1}{2}(221°) = 110.5°$.

5. $m\angle BAC + m\angle CAD = 95° + 64° = 159° = m\angle BAD = m\overarc{BD}$.

$m\angle BED = \frac{1}{2}m\overarc{BD} = \frac{1}{2}(159°) = 79.5°$.

$m\angle CAD = 64° = m\overarc{CD}$

$m\angle CED = \frac{1}{2}m\overarc{CD} = \frac{1}{2}(64°) = 32°$.

6. The arcs given total $47° + 102° + 74° + 43° = 266°$. $\overarc{NI}$ and $\overarc{JK}$ are congruent (because their chords are congruent) and comprise the remaining degrees in the circle.

So, $m\widehat{NI} + m\widehat{JK} = 360° - 266° = 94°$ and $m\widehat{NI} = m\widehat{JK} = \frac{1}{2}(94°) = 47°$.

$m\angle NMI = \frac{1}{2}m\widehat{NI} = \frac{1}{2}(47°) = 23.5°$. $m\angle JKL = \frac{1}{2}m\widehat{JK} = \frac{1}{2}m\widehat{NI} = m\angle NMI = 23.5°$.

$m\angle NIM = \frac{1}{2}m\widehat{NM} = \frac{1}{2}(47°) = 23.5°$. $m\angle LJK = \frac{1}{2}m\widehat{LK} = \frac{1}{2}(74°) = 37°$.

7. $m\angle YZW = \frac{1}{2}m\widehat{YW} = \frac{1}{2}(77°) = 38.5°$. $m\angle ZWY = \frac{1}{2}m\widehat{ZY} = \frac{1}{2}(160°) = 80°$.

$m\widehat{ZYW} = 160° + 77° = 237°$, so $m\widehat{ZW} = 360° - 237° = 123$.

$m\angle ZYW = \frac{1}{2}m\widehat{ZW} = \frac{1}{2}(123°) = 61.5°$.

8. $m\angle ZIK = \frac{1}{2}m\widehat{ZK} = \frac{1}{2}(85°) = 42.5°$. $m\angle KZJ + m\angle KIJ = \frac{1}{2}m\widehat{KIJ} + \frac{1}{2}m\widehat{KZJ}$

$m\angle KZI = \frac{1}{2}m\widehat{KI} = \frac{1}{2}(95°) = 47.5°$. $= \frac{1}{2}(95° + 120°) + \frac{1}{2}(85° + 60°)$

$m\angle ZIJ = \frac{1}{2}(60°) = 30°$. $= \frac{1}{2}(215°) + \frac{1}{2}(145°) = 107.5° + 72.5°$

$m\angle KZJ = m\angle KZI + m\angle IZJ = 47.5° + 60° = 107.5°$. $= 180°$

$m\angle ZJI = \frac{1}{2}m\widehat{ZKI}$ (*ZKI is a semicircle*) $m\angle IZJ = \frac{1}{2}m\widehat{IJ} = \frac{1}{2}(120°) = 60°$.

$\qquad = \frac{1}{2}(180°)$ $m\angle KIJ = m\angle KIZ + m\angle ZIJ = \frac{1}{2}m\widehat{KZ} + \frac{1}{2}m\widehat{ZJ}$

$\qquad = 90°$ $= \frac{1}{2}(85°) + \frac{1}{2}(60°)$

$m\angle ZKI = \frac{1}{2}m\widehat{ZJI} = \frac{1}{2}(180°) = 90°$ (*ZJI is a semicircle*). $= 42.5° + 30° = 72.5°$

$m\angle ZKI + m\angle ZJI = 180°$.

9. $m\widehat{CL} = m\angle CKL = 64°$. $m\widehat{LV} = m\angle LKV = 64°$.

$m\angle LUV = \frac{1}{2}m\widehat{LV} = \frac{1}{2}(64°) = 32°$. $m\widehat{LVU}$ is a semicircle so $m\widehat{LVU} = 180°$.

$m\widehat{LVU} = m\widehat{LV} + m\widehat{VG} + m\widehat{UG}$. $180° = 64° + 64° + m\widehat{UG}$, so $180° - 128° = m\widehat{UG} = 52°$.

$m\widehat{VU} = m\widehat{VG} + m\widehat{UG} = 64° + 52° = 116°$. $m\angle UVG = \frac{1}{2}m\widehat{UG} = \frac{1}{2}(52°) = 26°$.

$\angle KVG$ is not an inscribed angle, so we must think differently about it. ΔKUV is an isosceles triangle because the legs are radii, so $m\angle KUV = m\angle KVU = 32°$.
$m\angle KVU + m\angle UVG = m\angle KVG = 32° + 26° = 58°$.

Page 16
1. $m\widehat{JW} = 2m\angle JHW = 2 \cdot 60° = 120°$.

$m\widehat{JH} + m\widehat{JW} + m\widehat{HW} = 360° = 134° + 120° + m\widehat{HW}$, so $m\widehat{HW} = 106°$.

$m\angle JWH = \frac{1}{2}m\widehat{JH} = \frac{1}{2}(134°) = 67°$.
You can find $m\angle HJW$ either by the fact that there are 180° in a triangle or by the rule for inscribed angles. $m\angle HJW = 53°$.

2. $m\widehat{DSP} = 60° + 160° + 100° = 320°$. $m\widehat{DP} = 360° - 320° = 40°$.

$m\angle DSJ = \frac{1}{2}m\widehat{DJ} = \frac{1}{2}(100° + 40°) = \frac{1}{2}(140°) = 70°$.

$m\angle SDP = \frac{1}{2}m\widehat{SJP} = \frac{1}{2}(160° + 100°) = \frac{1}{2}(260°) = 130°$.

$m\angle DPJ = \frac{1}{2}m\widehat{DSJ} = \frac{1}{2}(60° + 160°) = \frac{1}{2}(220°) = 110°.$

$m\angle PJS = \frac{1}{2}m\widehat{SP} = \frac{1}{2}(60° + 40°) = \frac{1}{2}(100°) = 50°.$

3. $m\angle JIK = \frac{1}{2}m\widehat{JK} = \frac{1}{2}(50°) = 25°.$ $\quad\quad m\angle IJL = \frac{1}{2}m\widehat{IL} = \frac{1}{2}(100°) = 50°.$

$m\widehat{IJ} = 180° - m\widehat{JK} = 180° - 50° = 130°.$ $\quad m\angle IKJ = \frac{1}{2}m\widehat{IJ} = \frac{1}{2}(130°) = 65°.$

$m\widehat{LK} = 180° - m\widehat{IL} = 180° - 100° = 80°.$ $\quad m\angle LJK = \frac{1}{2}m\widehat{LK} = \frac{1}{2}(80°) = 40°.$

4. $m\widehat{EF} + m\widehat{FG} + m\widehat{GH} + m\widehat{HI} + m\widehat{IJ} + m\widehat{JE} = 360°.$ Each of these arcs is congruent as they have congruent chords. So, $6m\widehat{EF} = 360°$ and $m\widehat{EF} = 60°.$

$m\widehat{EGH} = 3m\widehat{EF} = 3 \cdot 60° = 180°$

$m\angle JIH = \frac{1}{2}m\widehat{JFH} = \frac{1}{2}(4 \cdot 60) = \frac{1}{2}(240) = 120°.$

5. $m\widehat{OF} = m\widehat{FD} = 43°.$ $\quad\quad m\widehat{OD} = m\widehat{OF} + m\widehat{FD} = 86°.$ $\quad m\angle OLD = \frac{1}{2}m\widehat{OD} = \frac{1}{2}(86°) = 43°.$

$m\angle LOD = \frac{1}{2}m\widehat{LD} = \frac{1}{2}(170°) = 85°.$ $\quad 180° - (43° + 85°) = 180° - 128° = 52° = m\angle ODL.$

$m\angle DRV = 90°.$

$m\angle RVD = 90° - 52° = 38°.$

Page 17

1. $4x = 3 \cdot 8 = 24,$ so $x = 6.$

$m\angle CED = 180° - 111° = 69°$

$\frac{1}{2}(m\widehat{AB} + m\widehat{CD}) = m\angle CED$

$m\widehat{AB} + 80 = 69 \cdot 2 = 138$

$m\widehat{AB} = 58$

$m\angle AED = \frac{1}{2}(m\widehat{AD} + m\widehat{BC})$

$111 = \frac{1}{2}(173° + m\widehat{BC})$

$222 = (173° + m\widehat{BC})$

$m\widehat{BC} = 49°$

$m\angle BEC = 111°$

$m\angle AEB = 69°$

2. $1 \cdot 9 = 2x,$ so $x = \frac{9}{2}$

$m\angle NKJ = 180° - m\angle MKJ$

$\quad\quad\quad = 180° - 120° = 60°$

$\frac{1}{2}(m\widehat{ML} + m\widehat{JN}) = \angle NKJ,$ so $\frac{1}{2}(m\widehat{ML} + 100°) = 60°$ and $m\widehat{ML} + 100° = 120°,$

therefore $m\widehat{ML} = 20°.$

$m\angle LKN = m\angle MKJ = 120°.$

$m\angle MKL = m\angle NKJ = 60°.$

$m\angle MKJ = \frac{1}{2}(m\widehat{MJ} + m\widehat{LN})$

$\quad\quad\quad = \frac{1}{2}(133° + 107°) = 120°$

3. $MO \cdot OT = NO \cdot OS$

$\quad 14x = 10 \cdot 24$

$\quad\quad x = \frac{120}{7} \approx 17.1$

$NQ \cdot QS = VQ \cdot QR$

$\quad 25 \cdot 9 = 18y$

$\quad\quad y = \frac{25}{2} = 12.5$

4. Since, $\overline{JV} \cong \overline{FM},$ $JV = 8;$ and, since $YC = 15,$ $FC = 15 - 7 = 8.$

$YJ = 15 + (15 - 8) = 15 + 7 = 22$

$YJ \cdot JV = LJ \cdot JH$

$22 \cdot 8 = 16JH$

$176 = 16JH$

$JH = 11$

$YF \cdot FV = MF \cdot FL$

$7 \cdot (15 + 8) = 8 \cdot FL$

$161 = 8FL$

$FL = \frac{161}{8}$

$m\widehat{HV} = 180° - (m\widehat{YM} + m\widehat{MH}) = 180° - (20° + 75°) = 180° - 95° = 85°.$

$m\angle MLH = \frac{1}{2}m\widehat{MH} = \frac{1}{2}(75°) = 37.5°.$

Page 18

1. $m\angle DHL = \frac{1}{2}m\widehat{DH} = \frac{1}{2}(70°) = 35°.$ $m\angle DHL + m\angle DHI = 180°$
$35° + m\angle DHI = 180°$
$m\angle DHI = 145°$

2. $m\angle SEQ = \frac{1}{2}m\widehat{SE} = \frac{1}{2}(42°) = 21°$ $m\angle KEN = \frac{1}{2}m\widehat{EK} = \frac{1}{2}(88°) = 44°.$
$m\angle SEQ + m\angle SEK + m\angle KEN = 180°$
$21° + m\angle SEK + 44° = 180°$
$65° + m\angle SEK = 180°$
$m\angle SEK = 115°.$

3. $m\angle ZXY = \frac{1}{2}m\widehat{XY} = \frac{1}{2}(118°) = 59°.$ $m\angle ZYX = \frac{1}{2}m\widehat{XY} = \frac{1}{2}(118°) = 59°.$
$m\angle ZXY + m\angle ZYX + m\angle XZY = 180° = 59° + 59° + m\angle XZY,$ so $m\angle XZY = 62°.$

4. $m\angle MCO = \frac{1}{2}m\widehat{CO} = 35°,$ so $m\widehat{CO} = 70°.$
$m\angle CPO = m\widehat{CO} = 70°.$

Page 20

1. 2 tangents. $360° - 200° = 160° = m\widehat{LR}.$
$\frac{1}{2}(m\widehat{LVR} - m\widehat{LR}) = m\angle Y = \frac{1}{2}(200° - 160°) = \frac{1}{2}(40°) = 20°.$

2. Tangent/secant. If $x = m\widehat{GM}$ and $360° - 84° = 276°,$ $276° - x = m\widehat{WVM},$
then $m\angle X = \frac{1}{2}(m\widehat{WVM} - m\widehat{GM}),$ so $53° = \frac{1}{2}((276° - x) - x),$ and $106° = 276° - 2x.$
Then, $2x = 170°,$ and $x = 85° = m\widehat{GM}.$

3. 2 secants. $m\widehat{IJ} - m\widehat{MJ} = m\widehat{MI},$ so $132° - 80° = 52° = m\widehat{MI}.$
$m\angle N = \frac{1}{2}(m\widehat{JA} - m\widehat{MI}),$ so $38° = \frac{1}{2}(m\widehat{JA} - 52°),$ and $76° = m\widehat{JA} - 52°,$ so $m\widehat{JA} = 128°.$
$m\widehat{JI} + m\widehat{IA} + m\widehat{JA} = 360° = 132° + m\widehat{IA} + 128°,$ so $m\widehat{IA} = 100°.$

4. Tangent/secant. $m\angle S = \frac{1}{2}(m\widehat{PQ} - m\widehat{UQ}),$ so $30° = \frac{1}{2}(139° - m\widehat{UQ}),$ and
$60° = 139° - m\widehat{UQ},$ so $m\widehat{UQ} = 79°.$
$m\widehat{PU} + m\widehat{UQ} + m\widehat{PQ} = 360° = m\widehat{PU} + 79° + 139°,$ so $m\widehat{PU} = 142°.$
$m\angle UPQ = \frac{1}{2}m\widehat{UQ} = \frac{1}{2}(79°) = 39.5°.$ $m\angle SQP = 180° - (30° + 39.5°) = 110.5°.$

5. $ED = EA + AD = 1.6 + 1.6 = 3.2\text{cm}.$ ΔEDG is a right triangle, so $3.2^2 + 2.4^2 = EG^2$ by the
Pythagorean Theorem. Therefore, $10.24 + 5.76 = EG^2 = 16$ and $EG = 4\text{cm}.$
$EJ = 4 - 1.5 = 2.5\text{cm}.$ $m\widehat{JD} = 180° - m\widehat{EJ} = 180° - 105° = 75°.$
$m\angle G = \frac{1}{2}(m\widehat{EBD} - m\widehat{JD}) = \frac{1}{2}(180° - 75°) = \frac{1}{2}(105°) = 52.5°.$
$m\angle JED = 90° - 52.5° = 37.5°.$

Page 21

1. $10^2 + 24^2$ ___ 26^2
   $100 + 576$ ___ 676
   $676 = 676$
   This is a tangent.

2. $4^2 + 4^2$ ___ 5^2
   $16 + 16$ ___ 25
   $32 > 25$
   This is not a tangent.

3. $4^2 + 7^2$ ___ 8^2
   $16 + 49$ ___ 64
   $65 > 64$
   This is not a tangent.

4. $YT = 3 + 2 = 5$
   $4^2 + 3^2$ ___ 5^2
   $25 = 25$
   This is a tangent.

Pages 23–24

1. $2.4^2 = 2(2 + x)$
   $5.76 = 4 + 2x$
   $1.76 = 2x$
   $.88\text{cm} = x$

2. $BU \cdot BL = BK \cdot BC$
   $16(16 + x) = 13(13 + 16)$
   $256 + 16x = 377$
   $16x = 121$
   $x = \frac{121}{16} \approx 7.6$

   $m \angle B = \frac{1}{2}(99° - 45°)$
   $\quad\;\; = \frac{1}{2}(54°)$
   $\quad\;\; = 27°$

3. $m \angle GDM = 90°$ because it is a right angle.

   $GD^2 + DM^2 = GM^2$
   $8^2 + 6^2 = GM^2$
   $64 + 36 = GM^2$
   $100 = GM^2$
   $10 = GM$

   $DM^2 = MV \cdot GM$
   $6^2 = (10 - x)10$
   $36 = 100 - 10x$
   $10x = 64$
   $x = 6.4$

   $\tan m\angle DGM = \frac{6}{8} = \frac{3}{4} = .75$
   $m\angle DGM \approx 36.9°$
   $m\angle DMG = 90° - m \angle DGM$
   $\quad\quad\quad\;\; = 90° - 36.9°$
   $\quad\quad\quad\;\; 53.1°$

4. $\angle P$ and $\angle G$ are congruent alternate interior angles formed by transversal $\overline{PG}$ cutting parallel segments $\overline{PF}$ and $\overline{XG}$. So, if $m\angle G = 24°$, then $m\angle P = 24°$. $m \angle PFO$ is a right angle, so $PF^2 + FO^2 = PO^2 = 25^2 + 11^2 = 746$ and $PO \approx 27.3$.
   $PO - 11 = PY = x = 16.3$.
   $m\widehat{XY} = 2\angle YGX = 2(24°) = 48°$.
   $m\widehat{FG} = 180° - m\widehat{YF} = 180° - 66° = 114°$.

   $m \angle FOP = 90° - 24° = 66°$.
   $m\widehat{YF} = m\angle YOF = 66°$.
   $m\widehat{GX} = 180° - m\widehat{YX} = 180° - 48° = 132°$.

5. $m\widehat{ZR} = m\widehat{ZH}$, because their chords are congruent. $m\widehat{ZR} = 113° = m\widehat{ZH}$.
   $m\widehat{HR} = 360° - (113° + 113°) = 134°$.　　$m\widehat{KR} = \frac{1}{2}m\widehat{HR} = 67°$.
   $m\angle RPH = \frac{1}{2}\left(m\widehat{RZH} - m\widehat{RH}\right) = \frac{1}{2}(226° - 134°) = 46°$.　　　$RP^2 = KP \cdot PZ$
   $ZP = ZY + YK + KP = 13 + 13 + 20 = 46$　　　　　　　　　　　$x^2 = 20(46)$
   $ZR = ZH = 21$　　　　　　　　　　　　　　　　　　　　　　　　$x^2 = 920$
   　　　　　　　　　　　　　　　　　　　　　　　　　　　　　　　$x = 2\sqrt{230} \approx 30.3$.

6. $LA + AW = LW = 1.6 + 3.8 = 5.4$.
   To find LS: $1.6 \cdot 5.4 = LS(LS + 2.5)$
   $8.64 = LS^2 + 2.5LS$
   　$0 = LS^2 + 2.5LS - 8.64$
   Using the Quadratic Formula, $a = 1$, $b = 2.5$, and $c = -8.64$.
   $$LS = \frac{-2.5 \pm \sqrt{(2.5)^2 - 4(1)(-8.64)}}{2(1)}$$
   So, $LC = 1.9 + 2.5 = 4.4$.
   $$= \frac{-2.5 \pm \sqrt{40.81}}{2} \approx 1.9.$$

   To find LK: $1.6 \cdot 5.4 = 5LK$, so $LK = \; \approx 1.7$. $KM = 5 - 1.7 = 3.3$.
   To find LG: $1.6 \cdot 5.4 = 1.7(1.7 + QG)$, so $\frac{8.64}{1.7} = 1.7 + QG \approx 5.1$, so $QG \approx 3.4$.
   Therefore, $LG = 3.4 + 1.7 = 5.1$.

7. $HP^2 + GP^2 = HG^2$ Note: This could also be done with $GZ \cdot GH = GP^2$.
$28^2 + 26^2 = HG^2$
$HG = \sqrt{1460} = 2\sqrt{365} \approx 38.2$ $m\angle G = \frac{1}{2}\left(m\widehat{HQP} - m\widehat{ZP}\right)$
$m\angle HPY = \frac{1}{2}m\widehat{HQ} = \frac{1}{2}(88°) = 44°.$ $= \frac{1}{2}(180° - 88°) = 46°.$
Since ΔHPY and ΔPHG share $\overline{HP}$ and have congruent corresponding angles, they are congruent triangles. $HY = GP = 26.$

8. $m\widehat{EN} = 180° - (51° + 109°) = 20°.$ $m\angle W = \frac{1}{2}\left(m\widehat{SK} - m\widehat{EN}\right) = \frac{1}{2}(51° - 20°) = 15.5°.$
$m\angle B = \frac{1}{2}\left(m\widehat{ENK} - m\widehat{SK}\right) = \frac{1}{2}(200° - 51°) = 74.5°.$ $KW = 13 + 13 + 14 = 40.$
$BW^2 = BK^2 + KW^2 = 121 + 1600 = 1721,$ so $BW = \sqrt{1721} \approx 41.5.$
$BW - BS = SW \approx 36.5.$ $WE \cdot WS = WN \cdot WK,$ so $36.5WE = 14 \cdot 40$ and $WE \approx 15.3.$

Page 25

1. First draw some auxiliary lines on the picture (the radii to the endpoints of the common tangent and a segment starting at the center of the smaller circle which is parallel to the common tangent). It helps if you break apart the picture and then apply the Pythagorean Theorem and the fact that opposite sides of a rectangle are congruent.
$1.5^2 = .5^2 + x^2$
$2.25 = .25 + x^2$
$x = \sqrt{2}$
$\quad \approx 1.4$

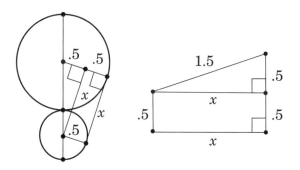

2. Externally tangent pair: $\odot S$ and $\odot J_{small}$.
Internally tangent pairs: $\odot I$ and $\odot J_{small}$, $\odot S$ and $\odot J_{large}$.

3. $YT = 1 = TW = UT.$ $YW = 2 \cdot YT = 2.$
$XW = .5 = XV.$ $WV = 2 \cdot XW = 2 \cdot .5 = 1.$
$ZV = ZW - WV = 3 - 1 = 2.$

4. $x^2 = 11 \cdot 25,\ x^2 = 275,$ so $x \approx 16.6.$ Using the right triangle: one leg is $14 + 7 = 21$, one leg is y and the hypotenuse is $y + 14$, so $21^2 + y^2 = (y + 14)^2$, therefore $441 + y^2 = y^2 + 28y + 196$. The y's cancel out, leaving $245 = 28y$, and $y = 8.75 \approx 9$.

Pages 26–27

1. We know that $EI = GI = 5$ and $IF = IH = 3$.
So $EI + IH = 5 + 3 = 8 = EH = GF.$

2. $\overline{NO} \cong \overline{PQ}$

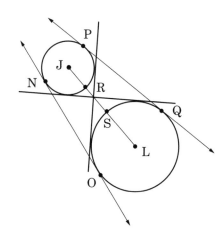

3. Common external tangent: $\overleftrightarrow{DE}$. There are no common internal tangents. $\overrightarrow{BD} = \overrightarrow{DE} = 4$ because if 2 segments are tangent to a circle from a single point, then they are congruent.

4. $DG = DE = 1.8$ and $EG = 2DG = 3.6$.
   $AE + ED + DG = 5$, so $AE + 1.8 + 1.8 = 5$ and $AE = 1.4$. ΔAFE and ΔDEF are both right triangles, because $\overleftrightarrow{EF}$ is a common internal tangent. Using the Pythagorean Theorem, $ED^2 + EF^2 = DF^2 = 1.8^2 + 1.6^2 = 3.24 + 2.56 = 5.8$, so $DF \approx 2.4$. Using the same thinking, $AE^2 + EF^2 = AF^2 = 1.4^2 + 1.6^2 = 1.96 + 2.56 = 4.52$ and $AF \approx 2.1$. From basic trigonometry, the tangent is a good choice for finding the angle measures.

   $\tan m\angle EFA = \frac{AE}{EF} = \frac{1.4}{1.6} = .875$, so $m\angle EFA \approx 41°$.

   $\tan m\angle EFD = \frac{ED}{EF} = \frac{1.8}{1.6} = 1.125$, so $m\angle EFD \approx 48°$.

   $m\angle AFD = m\angle EFA + m\angle EFD = 41° + 48° = 89°$.

5.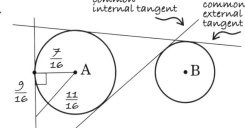

   $$\left(\frac{9}{16}\right)^2 + \left(\frac{7}{16}\right)^2 = \left(\frac{11}{16}\right)^2$$
   $$81 + 49 = 121$$
   $$130 \neq 121$$

   The line was not tangent.

Page 29

1. $A = \pi r^2 = \pi 3^2 = 9\pi \, \text{in}^2 \approx 28.3 \, \text{in}^2$. $C = 2\pi r = 2\pi 3 = 6\pi \, \text{in} \approx 18.8 \text{in}$.

2. $r = \frac{1}{2}d = \frac{1}{2}(8) = 4$. $A = \pi r^2 = \pi 4^2 = 16\pi \, \text{cm}^2 \approx 50.3 \, \text{cm}^2$. $C = \pi d = 8\pi \, \text{cm} \approx 25.1 \, \text{cm}$.

3. $A = \pi r^2 = \pi 5^2 = 25\pi$. $C = 2\pi r = 2\pi 5 = 10\pi$. When not dealing with units (inches, centimeters, etc.) it is common to leave answers in terms of pi.

4. $r = \frac{1}{2}d = \frac{1}{2}(25) = 12.5$. $A = \pi r^2 = \pi(12.5)^2 = 156.25\pi$. $C = \pi d = 25\pi$.

5. $KL \cdot LM = LI \cdot LG$, so $5x = 9 \cdot 16$ and $x = 28.8$. $KM = $ diameter $= 28.8 + 5 = 33.8$.
   Radius $= \frac{1}{2}(33.8) = 16.9$. $A = \pi r^2 = \pi(16.9)^2 = 285.61\pi \approx 897$. $C = \pi d = 33.8\pi \approx 106$.

6. $QW \cdot WC = WS^2$, so $15(15 + x) = 23^2$ and $225 + 15x = 529$, $15x = 304$, $x \approx 20.3 = $ diameter.
   Radius $= \frac{1}{2}(20.3) = 10.15$, so $A = \pi r^2 = \pi(10.15)^2 \approx 324$ and $C = \pi d = 20.3\pi \approx 64$.

Page 31

1. $A_{circle} = \pi r^2 = \pi(2.1)^2 \approx 13.85$.   This is an isosceles triangle. Drawing the altitude to the base, we are able to determine its height. $h^2 + 1.6^2 = 3.7^2$, so $h^2 = 13.69 - 2.56 = 11.13$ and $h \approx 3.34$. $A_{triangle} = \frac{1}{2}bh = \frac{1}{2}(3.2)(3.34) \approx 5.34$.
   The area of the shaded region is $13.85 - 5.34 \approx 8.5$.

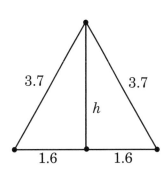

2. $A_{\frac{1}{4}circle} = \frac{1}{4}\pi r^2 = \frac{1}{4}\pi 8^2 \approx 16\pi \approx 50.3$.

3. This is a rectangle with a semicircular region. $A_{rectangle} = 24 \cdot 30 = 720$ and $A_{semicircle} = \frac{1}{2}\pi(12)^2 = 72\pi$, so $A_{shaded} = 720 + 72\pi \approx 946$.

4. $A_{rectangle} = (4.4)(2) = 8.8.$ $\qquad A_{circle} = \pi 1^2 = \pi.$ $\qquad 8.8 - \pi \approx 5.7.$

5. $A_{square} = 3^2 = 9.$ $\qquad A_{circle} = \pi(1.5)^2 = 2.25\pi.$ $\qquad A_{shaded} = 9 - 2.25\pi \approx 1.9.$

6. The radius of the large circle is 8, so the radius of the small circle is 4. $A_{circle} = \pi 4^2 = 16\pi$. Use what you know about 30–60–90 triangles and equilateral triangles to find the base and height of the triangle.
$A_{triangle} = \frac{1}{2}(8\sqrt{3})(12) = 48\sqrt{3}.$
$A_{shaded} = 48\sqrt{3} - 16\pi \approx 32.9.$

Page 34

1. $N = 180° - 120° = 60°$ and $r = \frac{1}{2}(26) = 13$.
$A = \frac{N}{360}\pi r^2 = \frac{60}{360}\pi 13^2 = \frac{1}{6}(169)\pi \approx 28.2\pi \approx 88.5$.

2. Although we don't usually think of central angles as being greater than 180°, for the purpose of this problem, the central angle of the shaded sector is $360° - 84° = 276°$. If you prefer to work with central angles which are less than 180°, you could find the area of the sector which is not shaded and then subtract that area from the total area in the circle. $N = 276°$ and $r = 10$, so $A = \frac{276}{360}\pi 10^2 = \frac{230}{3}\pi \approx 240.9$.

3. Since the inscribed angle has a measure of 47°, the intercepted arc has a measure of 94°, as does the central angle. $N = 94°$ and $r = 8$, so $A = \frac{94}{360}\pi 8^2 = \frac{752}{45}\pi \approx 52.5$.

4. $N = \frac{360}{3} = 120°$ and $r = 4$, so $A = \frac{120}{360}\pi 4^2 = \frac{16}{3}\pi \approx 16.8.$

5. There are 2 congruent sectors shaded. $N = 180° - 121° = 59°$ and $r = 9$.
$A = 2\left(\frac{59}{360}\pi 9^2\right) = \frac{531}{20}\pi \approx 83.4$

6. This is a square, so the diagonals form 90° central angles. The shaded regions aren't sectors, so we'll have to subtract the area of a triangle from the area of a sector and then take 2 of them because there are two shaded regions. The triangles are 45–45–90 triangles, so since the chords have lengths of $4\sqrt{2}$, the radii have lengths of 4.
$A_{sector} = \frac{90}{360}\pi 4^2 = 4\pi.$ $\quad A_{triangle} = \frac{1}{2}(4)(4) = 8.$ $\quad A_{2\ shaded\ regions} = 2(4\pi - 8) = 8\pi - 16 \approx 9.1.$

Page 36

1. Arc length $= x = \frac{m\overset{\frown}{ARC}}{360}2\pi r = \frac{144}{360}2\pi 5 = 4\pi \approx 12.6.$

2. The arc is $\frac{1}{5}$ of the circle, so $\frac{360}{5} = 72°$. $x = \frac{72}{360}2\pi 8 = \frac{16}{5}\pi \approx 10.1.$ $y = 2x = 2(10.1) = 20.2.$

3. $360° - 73° = 287°$, so $x = \frac{287}{360}2\pi 3 = \frac{287}{60}\pi \approx 15.0$.

4. $x = \frac{69}{360}2\pi 12 = \frac{23}{5}\pi \approx 14.5.$ $\qquad y = \frac{69}{360}2\pi 20 = \frac{23}{3}\pi \approx 24.1.$

5. $2.3cm = \frac{x}{360}2\pi 1.4cm$, so $\frac{2.3cm \cdot 360}{2\pi 1.4cm} = x \approx 94.1°$.

6. $z = 180° - 136° = 44°$. $\quad 1cm = \frac{44}{360}2\pi x$, so $x = \frac{360cm}{44 \cdot 2\pi} = \frac{45}{11\pi} \approx 1.3.$ $\quad y = \frac{180}{360}2\pi 1.3 \approx 4.1.$

Page 37

1. $(x-h)^2 + (y-k)^2 = r^2$, so $(x-(-3))^2+(y-0)^2 = 3^2$ and $(x+3)^2 + y^2 = 9$.

2. $(x-0)^2 + (y-0)^2 = 4^2$, so $x^2 + y^2 = 16$.

3. $(x-8)^2 + (y+4)^2 = 1^2$, so $(x-8)^2 + (y+4)^2 = 1$.

4. $r = \sqrt{(x_2 - x_1)^2 + (y_2 - y_1)^2} = \sqrt{(4-1)^2 + (4-2)^2} = \sqrt{3^2 + 2^2} = \sqrt{9+4} = \sqrt{13}$
$(x-1)^2 + (y-2)^2 = \sqrt{13}^2$, so $(x-1)^2 + (y-2)^2 = 13$.

5. $r = \sqrt{(2-(-1))^2 + (-2-(-7))^2} = \sqrt{3^2 + 5^2} = \sqrt{9+25} = \sqrt{34}$, so $(x+1)^2 + (y+7)^2 = 34$.

6. $r = \sqrt{(-4-2)^2 + (5-(-3))^2} = \sqrt{(-6)^2 + 8^2} = \sqrt{36+64} = \sqrt{100} = 10$
$(x-2)^2 + (y+3)^2 = 100$.

Page 39

1. $\odot A$ **2.** $\odot D$

3. $\odot H$ **4.** $\odot K$

5. Center $(2, 8)$,
radius $\sqrt{25} = 5$ $\odot P$

6. Center $(1, -4)$,
radius $\sqrt{4} = 2$ $\odot S$

7. Center $(-5, 0)$,
radius $\sqrt{49} = 7$ $\odot U$

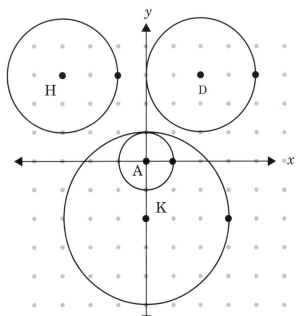

Each dot represents one unit.

8. Center $(0, 0)$, radius $\sqrt{1} = 1$, $\odot N$.
It is called the unit circle
because its radius has a length of
1 unit and its center is the origin
of the coordinate system.

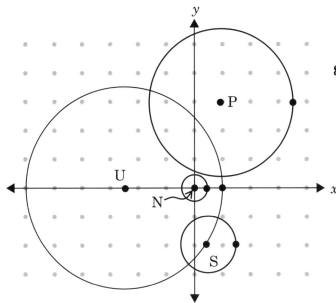

Each dot represents two units.

Pages 41–42 Circles Exam

1. $\overline{DC}$

2. $\overline{AD}$

3. $\overline{OD}$ or $\overline{OA}$

4. $\overleftrightarrow{AB}$

5. $\overline{DB}$

6. $m\angle ADB = \frac{1}{2}m\widehat{AB} = \frac{1}{2}(120^\circ) = 60^\circ.$

7. $m\angle AOB = m\widehat{AB} = 120^\circ.$

8. $m\widehat{ADB} = 360^\circ - m\widehat{AB} = 360^\circ - 120^\circ = 240^\circ.$

   So $m\angle ACB = \frac{1}{2}(m\widehat{ADB} - m\widehat{AB} = \frac{1}{2}(240^\circ - 120^\circ) = \frac{1}{2}(120^\circ) = 60^\circ.$

9. 90°

10. $m\angle AEB = \frac{1}{2}m\widehat{AB} = m\angle AFB = 20^\circ.$

11. $m\angle G = \frac{1}{2}(m\widehat{AB} - m\widehat{EF}) = \frac{1}{2}(95^\circ - 25^\circ) = \frac{1}{2}(70^\circ) = 35^\circ.$

12. $m\angle AHB = \frac{1}{2}(m\widehat{AB} + m\widehat{EF}) = \frac{1}{2}(70^\circ + 30^\circ) = \frac{1}{2}(100^\circ) = 50^\circ.$

13. $m\angle EHF = \frac{1}{2}(m\widehat{AB} + m\widehat{EF}) = 59^\circ = \frac{1}{2}(85^\circ + m\widehat{EF});$

    so, $118^\circ = 85^\circ + m\widehat{EF}$ and $m\widehat{EF} = 33^\circ.$

14. $x \cdot 5 = 3 \cdot 6,$ so $5x = 18$ and $x = \frac{18}{5} = 3.6.$

15. $x(x + 2) = 3(3 + 5).$ So $x^2 + 2x = 24$ and $x^2 + 2x - 24 = 0.$

    Therefore, $x = \frac{-2 \pm \sqrt{2^2 - 4(1)(-24)}}{2(1)} = \frac{-2 \pm \sqrt{100}}{2} \frac{-2 \pm 10}{2} = 4$ or $-6.$

    But x cannot equal a negative number in this situation, so $x = 4.$

16. $x^2 = 4(4 + 9) = 52,$ so $x = 2\sqrt{13} \approx 7.2.$

17. $5^2 + 12^2 = x^2,$ so $169 = x^2$ and $x = 13.$

18. $r = \frac{d}{2} = \frac{18}{2} = 9.$ $A = \pi 9^2 = 81\pi \approx 254.5.$

19. $C = 2\pi r = 22\pi,$ so $r = 11.$ $A = \pi 11^2 = 121\pi \approx 380 \, \text{cm}^2.$

20. $A = \frac{40}{360}\pi 8^2 = \frac{64}{9}\pi \approx 22.3.$

21. $A_{square} = 5^2 = 25.$

    $A_{\frac{1}{4}circle} = \frac{1}{4}\pi 5^2 = \frac{25}{4}\pi.$

    $A_{shaded\ region} = 25 - \frac{25}{4}\pi \approx 5.4.$

22. $20\pi = \frac{72}{360}\pi r^2,$ so $r^2 = 100$ and $r = 10.$

    Therefore, $\widehat{AB} = \frac{72}{360} \cdot 2\pi 10 = 4\pi \approx 12.6.$

STRAIGHT FORWARD Math Series

The Straight Forward Math Series

is systematic, first diagnosing skill levels, then *practice*, periodic *review*, and *testing*.

Blackline

GP-006 Addition
GP-012 Subtraction
GP-007 Multiplication
GP-013 Division
GP-039 Fractions
GP-083 Word Problems, Book 1
GP-042 Word Problems, Book 2

The Advanced Straight Forward Math Series

is a higher level system to diagnose, practice, review, and test skills.

Blackline

GP-015 Advanced Addition
GP-016 Advanced Subtraction
GP-017 Advanced Multiplication
GP-018 Advanced Division
GP-020 Advanced Decimals
GP-021 Advanced Fractions
GP-044 Mastery Tests
GP-025 Percent
GP-028 Pre-Algebra, Book 1
GP-029 Pre-Algebra, Book 2
GP-030 Pre-Geometry, Book 1
GP-031 Pre-Geometry, Book 2
GP-163 Pre-Algebra Companion

Upper Level Math Series

GP-104 Algebra, Book 1
GP-105 Algebra, Book 2
GP-045 Trigonometry
GP-054 Geometry
GP-053 Pre-Calculus
GP-064 Calculus AB, Vol. 1
GP-067 Calculus AB, Vol. 2

Math Pyramid Puzzles

Math Pyramid Puzzles
ISBN 978-1-9308-2062-3
GP-162
5 two-sided puzzles

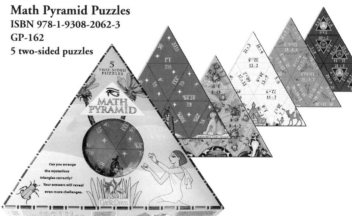

Assemble 5 two-sided puzzles each with different mathematical challenge. Solve the mathematical pyramid on the front side, turn the clear tray over to reveal of problem of logic: percents, decimals, fractions, exponents and factors.

Start building your pyramid at the bottom. The center piece is labeled and the picture may offer a clue.

Use your math skills to match sides with the same value.

You may find more than one match, but **all sides that touch** must match. When you are satisfied with your solution, close the tray.

Turn over and check the back. If the pieces are in order, you are correct!

Now, can you solve this logic puzzle?